WATER TREATMENT PLANT PERFORMANCE EVALUATIONS AND OPERATIONS

WATER TREATMENT PLANT PERFORMANCE EVALUATIONS AND OPERATIONS

JOHN T. O'CONNOR and TOM O'CONNOR
H_2O'C Engineering

RICK TWAIT
Superintendent of Water Purification, City of Bloomington, IL

Published by John Wiley & Sons, Inc., Hoboken, New Jersey
Published simultaneously in Canada

For general information on our other products and services or for technical support, please contact our Customer Care Department within the United States at (800) 762-2974, outside the United States at (317) 572-3993 or fax (317) 572-4002.

Wiley also publishes its books in a variety of electronic formats. Some content that appears in print may not be available in electronic formats. For more information about Wiley products, visit our web site at www.wiley.com.

Library of Congress Cataloging-in-Publication Data:

O'Connor, John, 1933 Feb. 11–
Water treatment plant performance evaluations and operations / John O'Connor, Tom O'Connor, Rick Twait.
p. cm.
ISBN 978-0-470-28861-0 (cloth)
1. Water treatment plants--Evaluation. 2. Water quality--Measurement. I. O'Connor, Tom, 1966– II. Twait, Rick. III. Title.
TD434.O278 2009
628.1'6--dc22 2008032179

Printed in the United States of America

10 9 8 7 6 5 4 3 2 1

CONTENTS

PREFACE

This book chronicles the development of advanced analytic laboratory capabilities at the Bloomington, Illinois water treatment plant over a 12-year period, 1997–2008. It details the application of these advanced analytical techniques to the evaluation of water treatment plant performance. Finally, it illustrates how the knowledge gained from these scientific evaluations can be used to improve the operations of the water treatment plant processes.

Initiated in 1997 as a cooperative effort between H_2O'C Engineering and the Bloomington (Illinois) Water Department operating and laboratory staff, a progressive series of water quality studies and plant process evaluations were conducted on-site at the Bloomington water treatment plant. This was part of an overall effort to effect cost savings, improve operational efficiencies, meet evolving regulatory requirements, update emergency procedures, and document operational procedures and experiences for the benefit of future generations of plant operators. Much of the material presented in this volume was derived from discussions with Bloomington's operators. The continual enhancement of Bloomington's water treatment laboratory capabilities and the presentation of highly focused in-house operator training programs have been by-products of this ongoing series of plant process evaluations.

A number of special in-house studies were undertaken, on a pilot scale and using Bloomington's lake water sources, to evaluate alternative treatment procedures for controlling tastes and odors. At significant cost, substantial amounts of data were obtained on the seasonal occurrence of compounds known to be associated with tastes and odors. Further, in an effort to define causative agents, numerous micrographs of microorganisms, from lake to tap water, were archived throughout a specific taste-and-odor episode.

The age, history, and unusual configuration of the Bloomington plant also allowed for a unique comparative evaluation of three generations of upflow, slurry blanket contact, and lime softener/clarifiers, manufactured successively by Dorr-Oliver, Infilco, and CB&I ClariCone.

Since the sand filters at Bloomington have been capped with granular activated carbon (GAC) for odor reduction, a series of special studies were conducted to determine means for improving the performance of the GAC in supporting microbial growth and reducing odor-causing compounds, such as geosmin and 2-methyl isoborneol (MIB). From the results, the seasonal timing of replacement of the GAC caps was modified. Finally, in-house pilot programs were initiated to assess the effectiveness of alternate odor control techniques, such as ozonation and ultraviolet light catalyzed peroxide oxidation.

CHAPTER 1 (MICROSCOPIC PARTICLE ANALYSIS)

After developing the necessary capabilities and defining the methodologies, epifluorescence (ultraviolet light) microscopy was used for the enumeration of *total bacterial cell counts.* This sensitive parameter was then used to evaluate the particle removal performances of coagulation and lime softening in Bloomington's three generations of upflow lime softener/clarifiers. In addition, cell counts were made following recarbonation, filtration, and disinfection. Thereafter, comparisons of microscopic techniques were made with turbidity and electronic particle counter measurements to assess filter backwash effectiveness and establish a more scientific basis for setting criteria for filter *return-to-service* following backwash. Finally, micrographs were taken to illustrate the appearance of the particles commonly found in Bloomington lake water sources and at various stages of treatment and distribution.

To facilitate the extended use of the microscopic procedure, Appendix A provides detailed procedures and instructions for conducting the *total bacterial cell count by epifluorescence microscopy.*

CHAPTER 2 (PLANT PROCESS EVALUATIONS)

More comprehensive studies were undertaken to assess the benefit of installing filter return-to-service *flow ramping* following backwash. Comparisons of filter effluent particle content were made using turbidity, direct microscopic enumeration, and electronic particle counts. While each measure is sensitive to different component parts of the spectrum of particles present in the filtered water, it was concluded that online monitoring of turbidity remains the most convenient and practical means for optimizing the backwash and flow ramping protocol for each individual filter.

CHAPTER 3 (LIME SOFTENING)

Studies conducted during the winter of 2000–2001 assessed the performance of Bloomington's lime softening process and determined the overall plant performance with respect to reduction of total organic matter (TOC). In addition, seasonal data on softening and the comparative removals of magnesium hydroxide and calcium carbonate were reviewed and compared with the results of jar test softening studies conducted over a wide range of pH. The significant effect of temperature on magnesium solubility and, therefore, coagulation by magnesium hydroxide is discussed with respect to its implication for operation during periods of cold weather.

CHAPTER 4 (ACIDIFICATION PROTOCOL)

Extensive in-house experimentation demonstrated that acidification of Bloomington softened water samples to pH 2.0 removed the finished water turbidity created by calcium carbonate but had no observable effect on the contribution to turbidity made by biotic particles. Therefore, filtered water sample acidification to dissolve inorganic precipitates increases the meaningfulness of turbidity as a microbiological surrogate. Based on these results, alternate turbidity exceedance levels were proposed and alternate operational procedures were established for routine filter performance evaluations.

CHAPTER 5 (FILTER OPERATIONS)

Based on extensive interviews with plant operators, an illustrated manual was prepared detailing filter operation, backwash procedures, performance monitoring, and filter maintenance.

Separately, as part of a *filter surveillance program*, the degree of filter bed expansion during backwash was measured using a shop-constructed pan flute device. The device and results are illustrated.

The removals of organic compounds on Bloomington's GAC-capped filters were quantified on both virgin and aged media to observe adsorption and microbiological uptake of dissolved organic carbon. Micrographs of the microbial growth on both GAC and sand media were archived. Operationally, the annual GAC and sand replacement procedures are illustrated.

The *Filter Operations Manual* details Bloomington's current filter washing procedures. Filtered water turbidities, both unacidified and acidified, are plotted for effluents during a complete filtration cycle. In addition, measurements of dissolved oxygen were used to assess the extent of bacterial respiration as well as their removal from filter media during backwash. From these studies, a novel procedure is advanced for evaluating biological activity on filter media.

Systematic efforts were undertaken to assess the cleaning of the filtration media as first one pump and then a second is started to progressively increase the degree of bed expansion. Observing insufficient expansion in older filters, recommendations are given for comparative testing of a replacement underdrain system with auxiliary air scour in a rehabilitated test filter.

CHAPTER 6 (GRIT REMOVAL)

The proprietary ClariCone, manufactured by CB&I, is an upflow, slurry blanket contact softener/clarifier that integrates the introduction of lime slurry, coagulation of precipitates within a slurry blanket, and settling to occur in an inverted cone-shaped basin.

The procedure for *gritting* (the process of removing the heaviest settled material from the lower cylinder of the softening unit) was evaluated by solids sampling and underwater video. In addition, the effect of the gritting process on the stability of the lime softening slurry blanket was evaluated. From these various evaluations, a revised gritting protocol was recommended.

To facilitate understanding of its operation, upward flow velocities in the inverted cone softener/clarifier were calculated and illustrated as a function of hydraulic loading and depth.

CHAPTER 7 (LIME SOFTENER PERFORMANCE ENHANCEMENTS)

The evaluation of a modified grit removal procedure was undertaken to minimize the disturbance of the lime softener *slurry blanket* during gritting and, thereby, mitigate problems created by *lime overfeed.*

Slurry blanket particle size profiles were obtained as a function of depth, both before and after gritting. As a result, the lime feed points to Bloomington's four inverted cone softeners were significantly modified to achieve more effective utilization of lime, minimize slurry blanket upset, and avoid transient lime overfeeds.

CHAPTER 8 (LIME SOFTENER OPERATIONAL ENHANCEMENTS)

The inverted cone softener receives its inflow through two pipes entering a cylindrical *can* tangentially at the base of the unit. The tangential inflows are intended to create a spiral flow pattern as the water rises within the cone. The distribution of the inflows between the two inlet pipes, one large and one small, is adjusted using two butterfly valves. This allows for control of the input kinetic energy required to promote an optimal spiral flow pattern. Too little energy input causes the spiral motion to *stall* so that the suspended matter in the blanket rises vertically. Such short-circuiting results in reduced residence time in the lime softening slurry blanket and causes surface upwellings or *boils* on the blanket surface.

To assist in operational control of the inflow valves, this evaluation explored the adjustment of flows to the inlet pipes for the establishment of an optimal blanket *swirl* (optimum hydraulic energy input). In addition, the studies aimed at determining the appropriate heights of the slurry blanket and the required setting of the height of the concentrator cone.

A discussion of the operation and maintenance of the complex slaked lime slurry delivery system is included in this chapter.

CHAPTER 9 (GRANULAR ACTIVATED CARBON)

Although taste-and-odor control was the primary goal when GAC caps were initially installed on Bloomington's 18 filters, other benefits realized included the supplemental reductions of TOC, pesticides, and herbicides (e.g., atrazine). This current reevaluation of the performance of the GAC included an assessment of the costs, required frequency of GAC replacement, warranties, and service contract conditions.

CHAPTER 10 (PLANT OPERATIONS MANUAL)

Photo- and micrograph-illustrated descriptions of Bloomington's water sources, watershed protection measures, treatment system (e.g., lime equipment and feed facilities, softening operations, recarbonation facilities, lime sludge disposal practice, filter operations, backwashing protocols, GAC replacement, primary and secondary disinfection, system control and data acquisition (SCADA) systems, laboratory and shop facilities, emergency power generator, and other backup systems) were derived from extensive interviews with plant operating and maintenance personnel, and plant design data. Constantly under revision, this document is being developed to serve as an introduction to plant configuration, nomenclature, and operation for newly recruited plant personnel.

The following series of brief, illustrated *guideline documents* (included in Appendix C) were also prepared to supplement the *Plant Operations Manual.*

Guideline Documents

1. Operator-on-Duty: Responsibilities
2. Operator's Laboratory: From Analysis to Database
3. Operator-on-Duty: *Making the Rounds*
4. The Lime Delivery System: From Dry Storage to Slakers to ClariCones
5. Recarbonation: From Liquid CO_2 to Gas to Solution
6. Filtration: Description and Operation

CHAPTER 11 (TASTE-AND-ODOR CONTROL)

Owing to a seasonal increase in nitrate levels in Lake Bloomington from agricultural drainage, on December 2, 2004, the influent to Bloomington's treatment plant was adjusted to a blend of 60% Lake Bloomington and 40% Evergreen Lake water. While effectively reducing nitrate levels, musty-earthy odors from Evergreen Lake water were detected in the plant's finished water within a day.

System assessments included an evaluation of nutrient input sources; microscopic examination and quantification of organisms through the plant; monitoring of geosmin, MIB, TOC, H_2S and tannin/lignin; and observation of lake stratification conditions, including temperature and oxygen profiles.

Preliminary assessments of responses to treatment alternatives (powdered activated carbon, GAC caps, aeration, potassium permanganate, chlorine dioxide, hydrogen peroxide, Fenton's reagent, ozone, aerobic biodegradation on GAC) were undertaken as an operational guide to the most effective and practical methods for taste-and-odor control.

A series of microscopic surveys of particle abundance provided micrographs of organisms from lake to tap water on numerous dates throughout the duration of the taste-and-odor episode.

CHAPTER 12 (GAC ADSORPTION AND MICROBIAL DEGRADATION)

A major objective of this testing program was to determine the effectiveness of organism growth on aged (used or in-service) GAC in providing biologically mediated removal of odorous compounds after the initial adsorptive capacity of the virgin GAC was exhausted. In addition to odor surrogate (geosmin) removal, a temperature-dependent dissolved oxygen depletion on microbially colonized GAC was observed.

Monitoring of geosmin and MIB as a function of lake water depth revealed that these compounds were most abundant just above the lake benthos.

Aeration and *ozonation* were compared with respect to geosmin odor removal. While effective, ozonation was found, instead, to impart an ozonous odor.

A special series of tests was made using a feed of hydrogen peroxide to plant filtered water followed by *ultraviolet irradiation* to catalyze the oxidation of geosmin. Preliminary results indicated that high peroxide dosages and extended UV contact times would be required for effective geosmin removal.

PROCESS EVALUATIONS AND OPERATIONS

Water supply, treatment, and distribution systems constantly evolve to meet increasing demands and address more stringent drinking water quality criteria. Continuing development of additional, new water sources often change water quality input parameters. However, improved analytical techniques provide better means for optimizing water treatment processes and achieving economies through improved operations.

Overall, external changes and developments create challenges to those methods previously employed for system operations. The authors hope that the scientific and technical methodologies illustrated in this case study will provide specific examples of how a progressive water utility can work to meet these challenges. In the spirit of contributing to these efforts to improve water treatment practice, questions, comments, and suggestions may be sent to john@h2oc.com.

Columbia, Missouri

Bloomington, Illinois
January, 2009

JOHN O'CONNOR
TOM O'CONNOR
RICK TWAIT

1

MICROSCOPIC PARTICLE ANALYSIS

Chapter 1 chronicles the beginning of what has been an independent, 12-year journey to improve water treatment plant process performance and operations at the Bloomington, Illinois water treatment plant. A partnership of engineers, operators, laboratory staff, and plant supervisors began this odyssey with the purchase of a modern incarnation of a venerable piece of scientific equipment, a light microscope. The creative use of this microscope plus the scientific apparatus subsequently acquired provided numerous opportunities for all who took part in this on-site effort to obtain greater insight into the actual working of their water treatment processes. With improved knowledge, this has led to improvements in equipment configurations, operational protocols, and cost efficiencies. The following descriptions of this work are intended to serve as guidelines for a more scientific approach to process evaluation and operation for the new generation of water utility operators and supervisors. In pursuing better scientific understanding of their own unique treatment processes and operations, the authors believe that water plant operation can become an ever more satisfying and dynamic professional career.

Microscopic evaluations of *source water quality*, water treatment plant *particle removal efficiency*, *lime softener performance*, and *filter backwash* were conducted as part of a program of integrating advanced microscopic capabilities into the Water Department laboratory. Figure 1-1 shows the compound microscope with epifluorescent attachment, video camera, and monitor utilized at Bloomington.

Since the Bloomington water treatment plant (Fig. 1-2) draws from lake water sources, epifluorescence (ultraviolet light) microscopy was adopted to evaluate the seasonal effectiveness of chemical coagulation and precipitative lime softening (Fig. 1-3) in removing source water particles, including bacteria, the most abundant micrometer-sized biotic particles.

Water Treatment Plant Performance Evaluations and Operations. By John T. O'Connor, Tom O'Connor, and Rick Twait

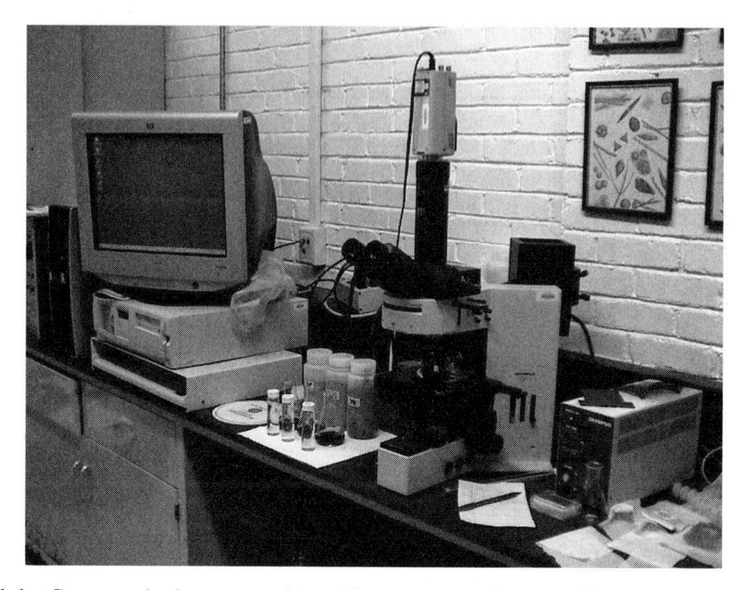

Figure 1-1 Compound microscope with epifluorescent attachment, video camera, and monitor.

Subsequently, in assessing filter backwash effectiveness and the performance of filters on *return-to-service*, microscopic techniques were compared with both turbidity and electronic particle counting measurements.

From these experiences, a simplified laboratory procedure for rapidly conducting the *total bacterial cell count by epifluorescence microscopy* was defined and is presented in Appendix A.

Figure 1-2 Treatment facilities at Lake Bloomington.

Figure 1-3 Lime softening units: external and housed.

BLOOMINGTON WATER SOURCES AND TREATMENT PLANT PROCESSES

Source Waters

Lake Bloomington and Evergreen Lake contain a large and diverse population of algae and bacterial cells (Figs. 1-3a–j; see color insert) that varies seasonally throughout the year.

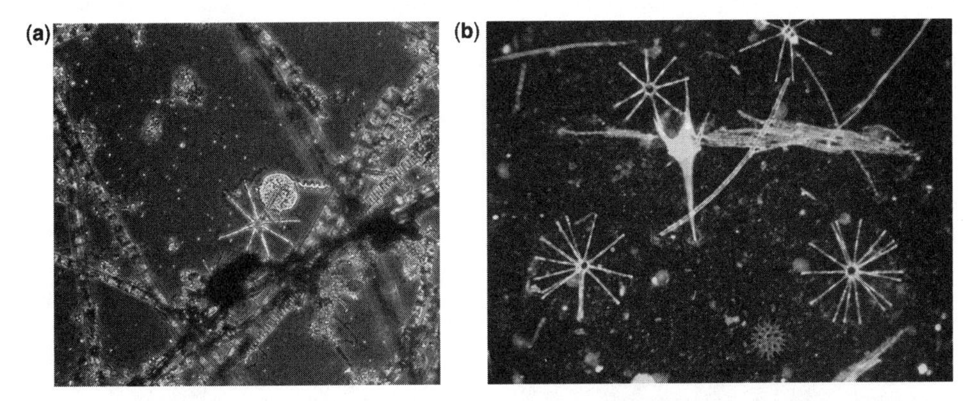

Figure 1-3a–b (a) Filamentous algae, diatoms, bacteria, and organic debris in Lake Bloomington water. Micrograph illustrates the wide distribution of particle sizes and shapes in lake water. Large, fluorescing particles are algae and diatoms (10 to 40 μm); smaller fluorescing dots are, primarily, bacterial cells (1 μm); (b) Micrograph of Lake Bloomington water reveals fragile diatoms (Asterionella), algal filaments, numerous bacterial cells, and loose particle aggregations. The flocculent extracellular excretions of microorganisms may entrain numerous smaller particles. Loose and fragile aggregations may be disrupted by passage through capillaries of particle counting devices. (See color insert.)

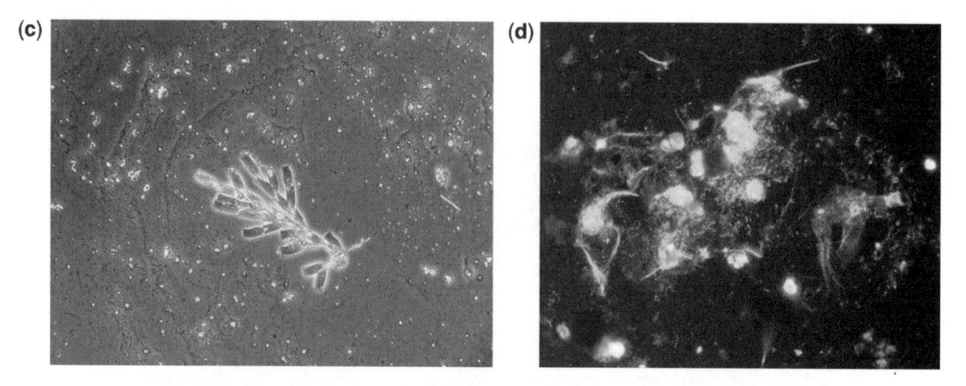

Figure 1-3c–d (c) Dinobryon colony in lake water; (d) Disintegrating *Daphnia* in lake water colonized by numerous bacterial cells. (See color insert.)

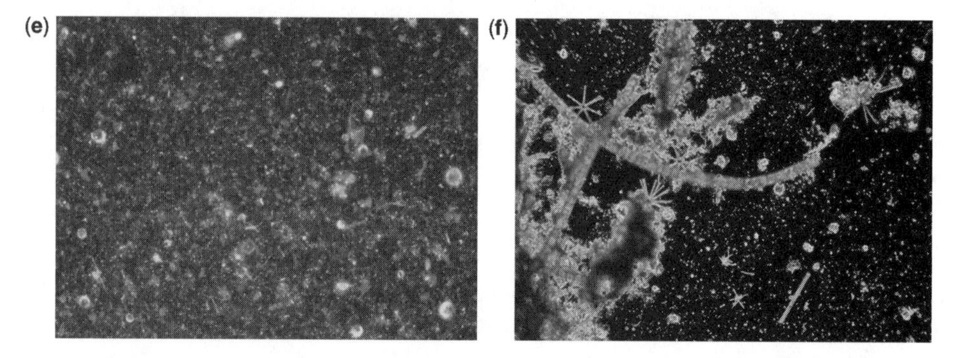

Figure 1-3e–f (e) Turbid Evergreen Lake water with inorganic solids and abundance of bacterial cells; (f) Evergreen Lake water with aggregations of debris attached to algal filaments. Smaller particles appear to be flocculated by extracellular polymers secreted by algae. (See color insert.)

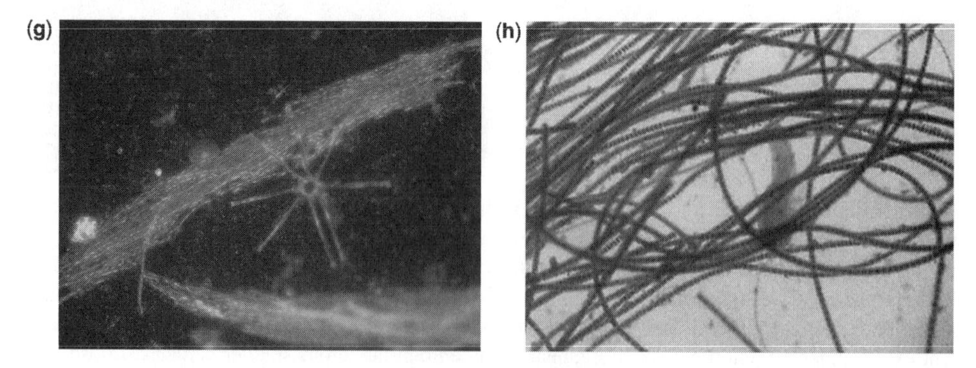

Figure 1-3g–h (g) Sheaths of filaments contain large numbers of algal cells; (h) Mat of algal filaments entrain grazing zooplankter. (See color insert.)

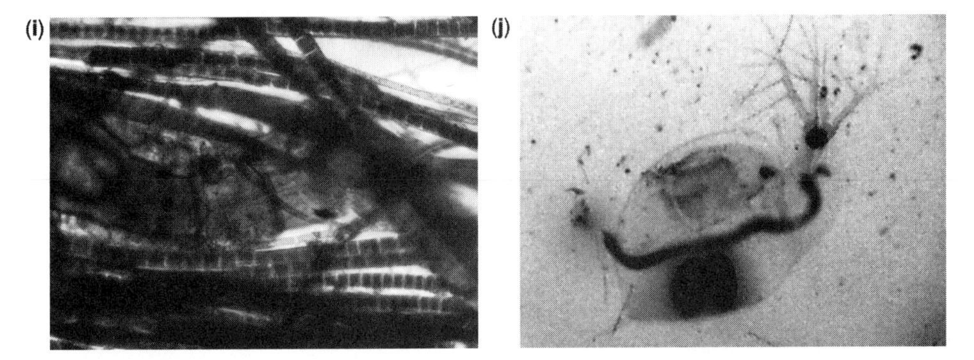

Figure 1-3i–j (i) Algal filaments showing well-defined cells enclose macroinvertebrate (j) Fecal matter in gut of *Daphnia*. Small, motile aquatic crustaceans with entrained microscopic organisms passing through treatment constitute *particles of potential health significance*. (See color insert.)

Inorganic and organic particulate matter may be abundant or minimal depending on the water temperature, lake level, degree of mixing, precipitation (stream inflow), wind, artificial mechanical destratification, and, even, powerboat traffic.

Plant Influent (Raw) Water

The flow diagram of Fig. 1-4 illustrates the sources and components of the Bloomington water treatment system. Initially, the lake water influents to the Bloomington water

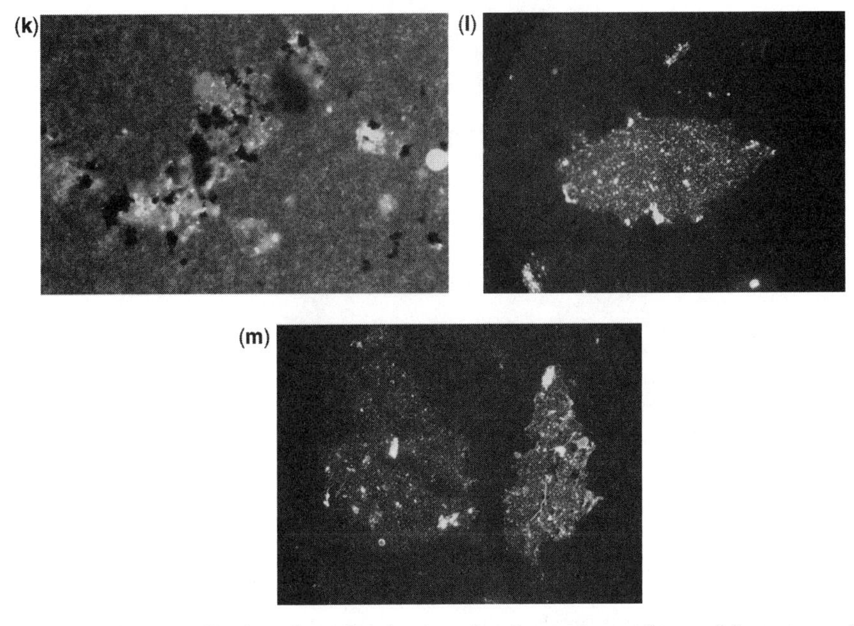

Figure 1-3k–m (k) Application of powdered activated carbon and coagulants to lake water results in formation of flocs with embedded carbon (black) and microorganisms (yellow); (l) Well-defined floc formed in softener shows near complete entrainment of micrometer-sized particles; (m) Fragile flocs in softener effluent show entrained bacteria, algal filaments, inorganic debris, and precipitates formed during softening in a matrix of magnesium hydroxide. (See color insert.)

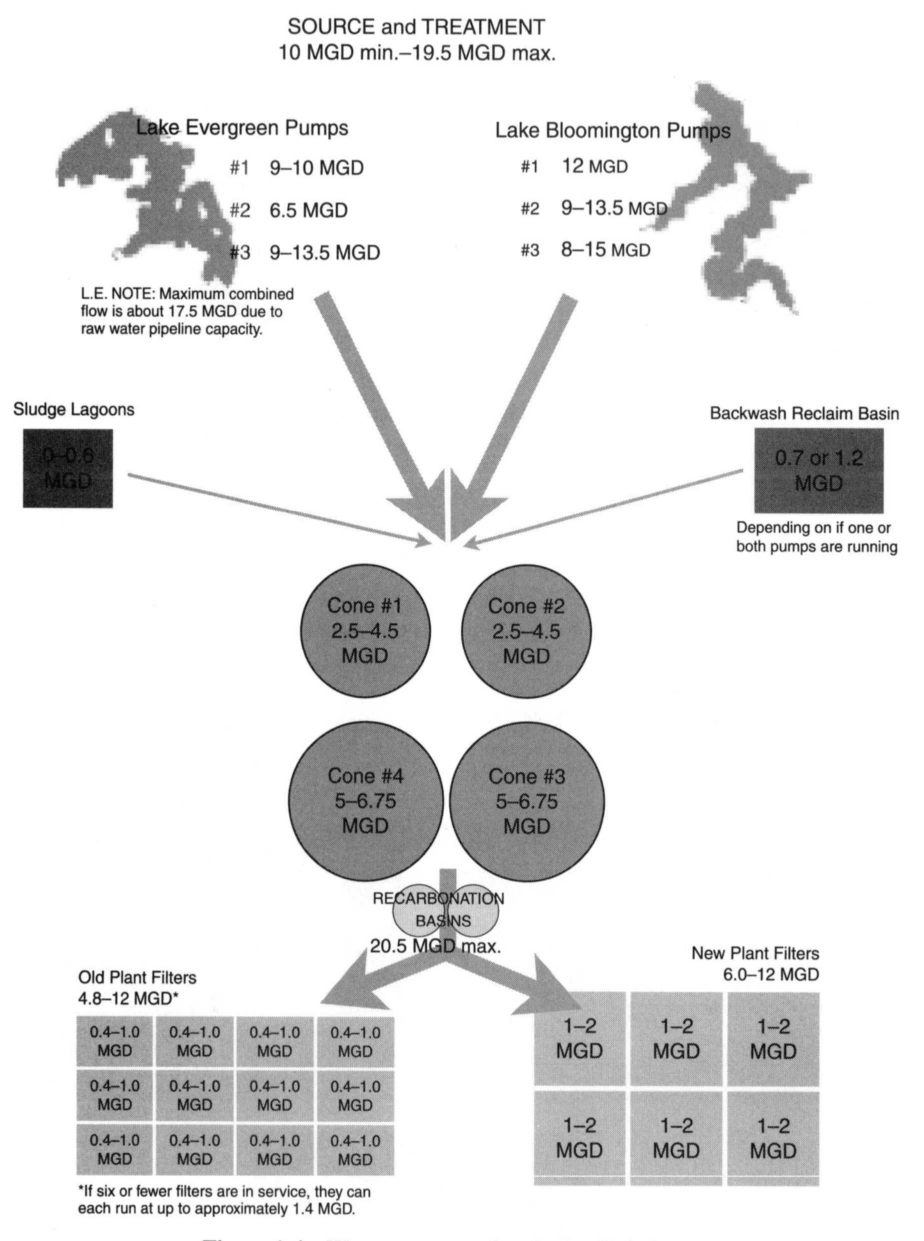

Figure 1-4 Water treatment plant hydraulic balance.

treatment plant are treated with a cationic polyelectrolyte upon withdrawal at the source. Polymer additions at the lake intakes allow for mixing during transmission that results in a degree of flocculation of microscopic lake water particles by the time the water reaches the treatment plant. While total organism counts may remain the same, their degree of aggregation increases.

Lime-Treated and Settled Water

The Bloomington Water Treatment Plant provides water softening by precipitation of calcium carbonate and magnesium hydroxide (at pH >11) with lime. Lime softening, even after effective settling, produces a settled effluent that contains microfloc with numerous particles and microorganisms embedded (Figs. 1-3k–m; see color insert).

Since three generations of lime softeners were in parallel operation when this evaluation was undertaken, a unique comparison could be made of their relative efficiency in microscopic particle removals. The newest and most efficient softening unit is shown in Fig. 1-5.

The settled solids (lime softening sludges) from the precipitation of hardness are transferred to a series of storage lagoons where the solids are dewatered (Fig. 1-6). The supernatants from the lagoons contain aged calcium carbonate particles. This high pH supernatant is periodically recovered (pumped) to the *reclaim basin* and intermittently mixed with filter backwash water. After settling, this blend is transferred to the plant influent at the rapid mix basin for recovery. Once more, most of the particles present in this reclaimed water consist of either freshly precipitated or aged calcium carbonate.

Recarbonated Water; Filter Influent

Softened and settled water from Bloomington's upflow slurry contact clarifiers passes into two recarbonation basins (Fig. 1-7). The addition of carbon dioxide gas within these basins lowers the pH, arrests further precipitation of calcium carbonate (stabilization), and may dissolve some of the newly precipitated calcium carbonate, thereby reducing inorganic particle concentrations and numbers before filtration. Figure 1-7a illustrates the appearance of particles in the recarbonation basin effluent which are also the filter influents.

The present microscopic evaluation has also shown that the high pH produced by the addition of lime does not cause problems with the enumeration of bacterial cells by epifluorescence microscopy. In addition, high concentrations of suspended organic

Figure 1-5 Upflow lime slurry contact softener/clarifier.

Figure 1-6 Lime softening sludge storage lagoons.

matter in the source water did not combine with the fluorochrome (acridine orange) to create high background fluorescence. High background fluorescence could make tiny, stressed bacterial cells difficult or impossible to enumerate.

Filter Effluent

The particles in the filter effluent of greatest concern to utility managers are those that may have an impact on the health of water consumers. While many of the particles in filtered

Figure 1-7 Recarbonation basin with serpentine effluent launders.

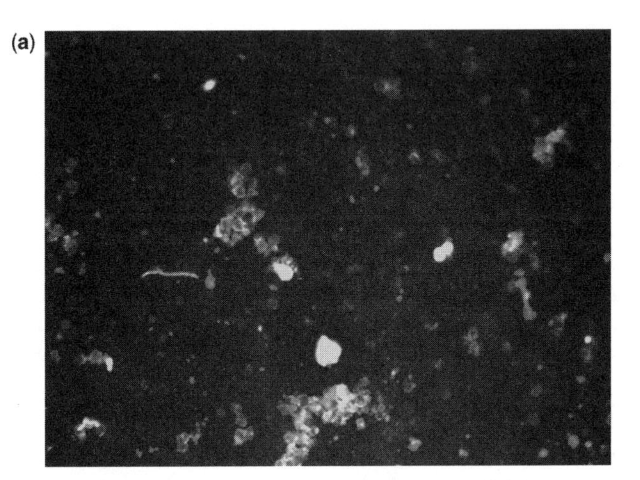

Figure 1-7a Calcium carbonate crystals dominate particles found in the recarbonation basin and, hence, filter influent. (See color insert.)

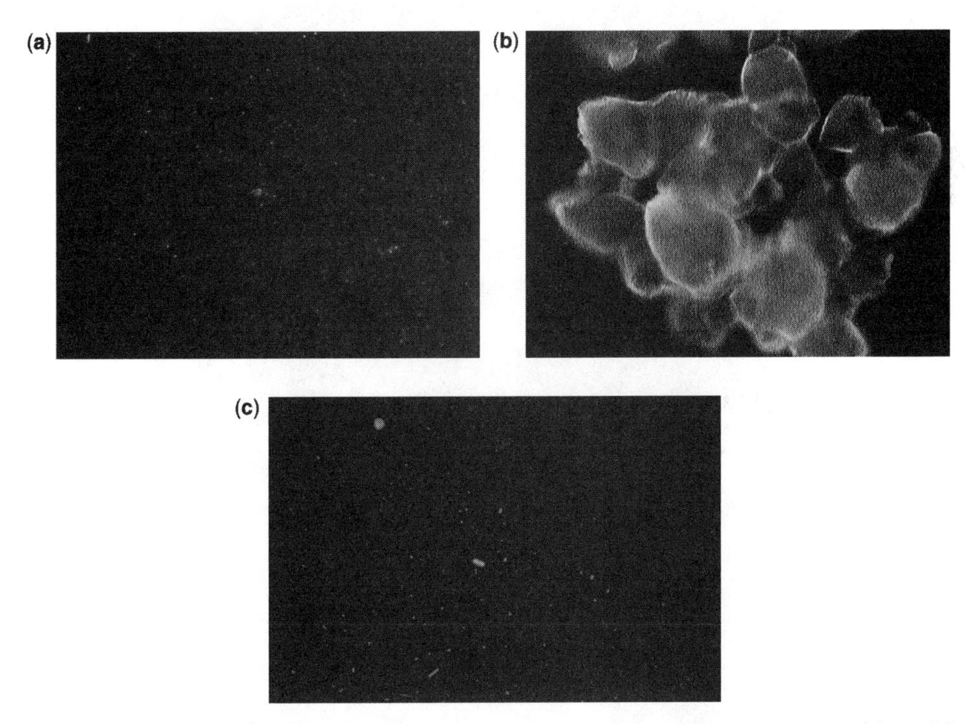

Figure 1-8a–c (a) Planktonic bacterial cells predominate in filter effluent. Unprotected bacterial cells are believed to be most susceptible to action of disinfectants; (b) Cluster of calcium carbonate crystals are occasionally found in filter effluent. Crystals in filtered water do not appear to harbor microorganisms; (c) Discolored (stressed) bacterial cells observed in filtered water following final disinfection. (See color insert.)

water may be calcium carbonate, rust, and innocuous organic and inorganic debris, those harboring potential pathogens may be designated as *particles of potential health significance*.

Major particles of concern include 4 to 10 μm biotic particles representative of the size range in which the cysts of *Giardia* and the oocysts of *Cryptosporidium* are found. In addition, bacteria in clumps or clusters large enough to interfere with the subsequent chemical inactivation (disinfection) of the cells are of special interest. Typically, as Fig. 1-8a illustrates, these larger particles and clumps were not observed in the filter effluents. On those occasions when particles were observed in finished water, they often appeared as clusters of calcium carbonate crystals, as shown in Fig. 1-8b. Finally, Fig. 1-8c shows the appearance of the filtered water following final disinfection.

The Bloomington Water Department, in the interest of improving water quality and controlling seasonal taste-and-odor problems, has installed *granular activated carbon* (GAC) atop its filters. Therefore, there is an interest in determining whether the abrasion of the GAC granules during backwash results in the liberation of *carbon fines* that may subsequently penetrate the filters and pass into the plant finished water. Based on microscopic particle enumeration data, treatment process adjustments could be made to minimize the passage of such particles.

Figure 1-8 shows the influent to the plant's oldest (1929) filters, and Fig. 1-9 is a schematic of the flow diagram and chemical application points.

Figure 1-8 Softened, settled, and recarbonated filter influent.

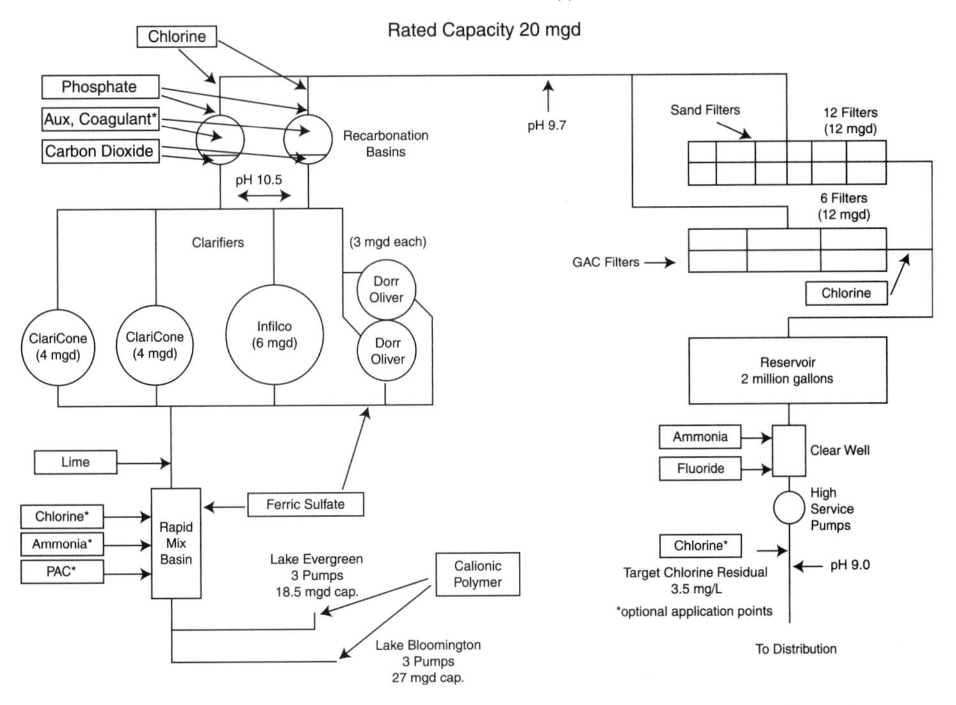

Figure 1-9 Water treatment plant schematic: chemical application points.

NUMBER OF BACTERIAL CELLS IN NATURAL AND TREATED WATERS

The observed number of bacterial cells in natural water sources ranges over orders of magnitude, from more than 10^6 cells/ml in surface waters to less than 10^3 cells/ml in deep well waters. Treated drinking waters commonly contain between 10^3 and 10^6 cells per milliliter, largely depending on water source and bacterial cell removal efficiency. Conventional surface water treatment processes employing coagulation, sedimentation, and filtration have been observed to obtain 90% cell removals. Removals are less during periods of low water temperature while during warm water periods, bacterial cell removals may increase to 99% or greater. Where settling is provided, most bacterial cell removal appears to be accomplished as a result of coagulation and sedimentation rather than filtration.

RESULTS OF MICROSCOPIC ENUMERATION

Bacterial Populations in Lake Bloomington Water

In 1997, Lake Bloomington water was sampled at intake depths of 0.3 and 6 meters. Bacterial cell counts were 2.3 and 1.8 million cells per milliliter, respectively, averaging about 2 million cells per milliliter. At the rapid mix sampling point, a comparable 1.8 million cells per milliliter were enumerated.

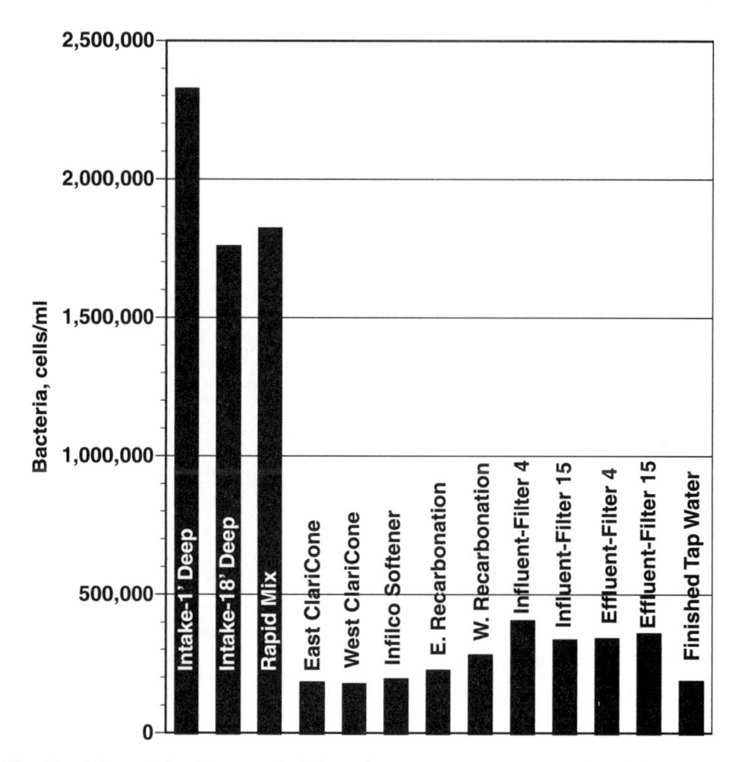

Figure 1-10 Total bacterial cell counts in Bloomington source, processed, and finished waters, 1997.

These total bacterial populations fall within the range reported for lake waters. However, the size, population, and proportion of bacteria attached to suspended particles (*particle-associated bacteria*) can be expected to vary greatly throughout the year. During periods of warm water temperature and rapid bacterial growth, most of the bacteria may be *planktonic*. Alternately, during cold temperature periods, bacteria may be primarily attached to other, larger particles. While this attachment will tend to improve removal of bacterial cells, overall removals may still deteriorate because of temperature-related decreases in chemical and physical treatment process rates and efficiencies (precipitation, coagulation, mixing, settling, filtration).

Figure 1-10 shows the results of enumerating total bacterial cell counts in Bloomington's source, processed, and finished waters.

Removal of Bacterial Cells during Lime Softening and Clarification

The effluents from Bloomington's array of lime softener/clarifiers (East ClariCone, West ClariCone, and Infilco units) exhibited 178,000, 173,000, and 191,000 cells per milliliter, respectively (Fig. 1-10). These relatively consistent results averaged 181,000 cells per milliliter and indicated an overall 91% reduction in total bacterial cell numbers by lime precipitation, contact with the slurry blankets, and sedimentation. Although not evident from microscopic observation, many of the remaining 9% of the cells may have been injured or inactivated due to the high pH (>11) attained by the addition of lime.

From these data, it appears that combined lime softening and clarification achieve substantial *physical* reductions in the numbers of bacteria initially present in the Lake Bloomington water source.

Recarbonation Basins

Following lime treatment and clarification, the softened water is recarbonated using carbon dioxide gas to reduce pH to ≈ 9.7 to arrest further precipitation of calcium carbonate. The bacterial cell populations observed in the two recarbonation basins were 221,000 and 277,000 cells per milliliter. Although higher than the numbers observed in the settled water, these populations appear comparable.

Influent to Filters

The influents to Filters 4 (an older unit) and 15 (a new unit) were found to contain 402,000 and 333,000 cells per milliliter, respectively. The average of these counts (368,000 cells/ml) is roughly twice the number observed in the softener/clarifier effluents. Only continued, repetitive monitoring may indicate whether this apparent change resulted from variations in influent bacterial populations, organism regrowth during recarbonation, or recruitment of cells from the walls of tanks and filter influent piping.

Effluent from Filters

The effluents from Filters 4 and 15 were found to contain 338,000 and 356,000 cells per milliliter, respectively, averaging 347,000 cells per milliliter. From this initial sampling, it would appear that most of the bacterial cells in the influent water were passing through the filters. This may be an indication that the bacteria were not embedded in a coagulant floc of a size that could be effectively retained by the filter media. Alternately, influent cell removals during filtration may have been offset by the contribution (recruitment) of new cells stripped from filter media. Either way, the cells observed in the filter effluent were predominately *planktonic*. It is believed that such unprotected, planktonic cells are those most readily inactivated by subsequent disinfection.

Finished (Disinfected) Water

Following disinfection, the finished tap water at the Bloomington Treatment plant exhibited 183,000 cells per milliliter, or roughly half the number enumerated in the filtered water. This apparent decrease may have been due to the progressive *lysing* (disintegration) of the injured cells following the application of disinfectant. If so, longitudinal (timed) studies of the bacterial cell count following disinfection may show a progressive decrease in the bacterial cell count.

Overall Bacterial Cell Reductions

Overall, a 90% reduction in the number of total bacterial cells was observed during treatment. As expected, most of the physical removal of these cells took place during softening and sedimentation.

Micrographs taken at various treatment stages are shown on Figs. 1-3a–m, 1-7a, and 1-8a–c. These illustrate the appearance of particles in the influent, the formation of calcium carbonate, and the removal of particles by sedimentation and filtration. Finally, the appearance of stressed cells, following disinfection, is shown in the finished water.

In evaluating treatment plant process efficiency with respect to the physical removal of micrometer-sized biotic particles, such as bacteria, this enumeration procedure must be routinely replicated in order to determine the effects of seasonal temperature changes on influent organism populations or the effect of modified treatment plant operations.

COMPARATIVE EVALUATION OF LIME SOFTENER/CLARIFIERS

Over a period of more than half a century, three generations of lime softening treatment units had been constructed and were operating in parallel at the Bloomington plant. This unusual circumstance made it possible to conduct a separate, comparative study of the performance of three progressively newer softener/clarifiers (Dorr-Oliver, Infilco Accelator, CB&I ClariCone) with respect to total bacterial cell removals. While all performed well with respect to softening and turbidity reductions, the cell counts obtained indicated that the newest units were achieving the most complete removal of bacterial cells (Fig. 1-11).

EFFECT OF BACKWASH ON FILTER PERFORMANCE

Bacterial Populations in Filter Effluent before and after Filter Backwash

A preliminary study was conducted of bacterial populations in a filter effluent before and after filter backwash. At the end of a normal, 48 hour filter run, and before backwash, the filter effluent bacterial population was found to be 776,000 cells per milliliter. Immediately following backwash, the bacterial cell count had reduced to 583,000 cells per milliliter. Over a period of the next hour, the population continued to decline to 338,000 cells per milliliter, or roughly 44% of the population observed before backwash.

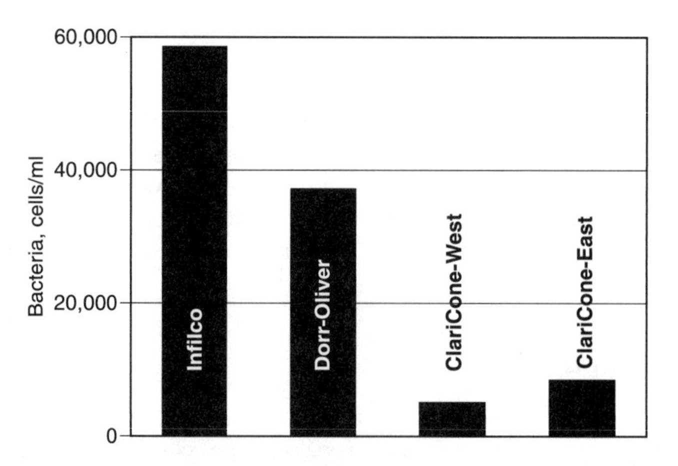

Figure 1-11 Total bacterial cell counts in effluents from lime softener/clarifiers.

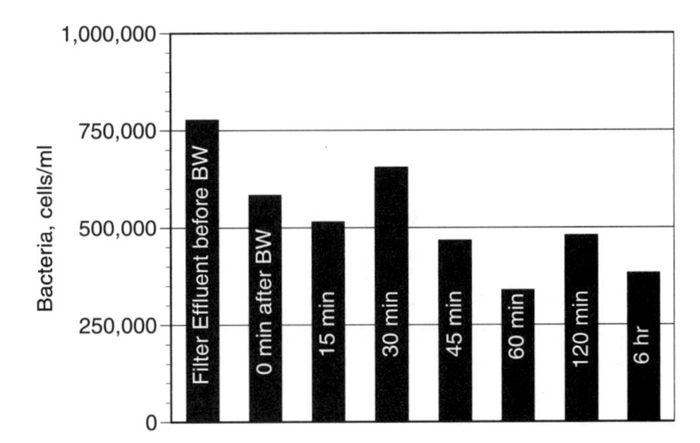

Figure 1-12 Total bacterial cell counts in filter effluents before and after filter backwash.

Since no significant peak was observed following backwash (Fig. 1-12), this initial study indicated that *filtering-to-waste* would not have significantly reduced the number of micrometer-sized biotic particles (bacteria) discharged to the finished water from this filter.

Turbidity of Filter Effluent before and after Backwash

Each filter unit at the Bloomington Plant is equipped with a continuous flow, low range turbidimeter for operational control of the filters. These continuous monitors indicate increases in turbidity through the filter cycle and warn when filters may require cleaning before scheduled backwash. They also monitor turbidity after filters return to service following backwash.

Turbidity monitoring data for a filter effluent, before and after backwash, is shown in Fig. 1-13. The results indicate that 15 minutes after backwash, filter effluent turbidity (0.13 ntu) had declined to one-half the turbidity measured before backwash. Turbidity continued to decline to 0.06 ntu over the next two hours.

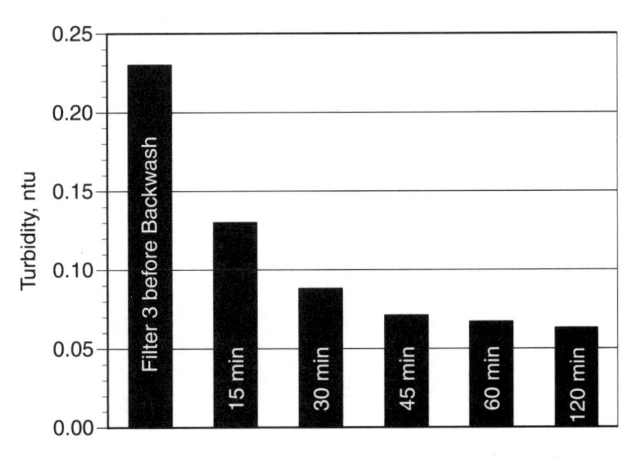

Figure 1-13 Turbidity of sand filter effluents before and after filter backwash.

These turbidity results appear consistent with the results from the measurement of total bacterial cell counts before and after backwash. While this may be an indication that filter-to-waste is not beneficial, measurements were not made prior to 15 minutes following backwash during which period a transient in turbidity may have occurred.

Monitoring of Filter Effluent Turbidity Following Backwash

A more detailed longitudinal assessment was made of another filter following backwash. This filter was newly returned to service after placement of a fresh 0.48 m deep granular activated carbon cap over a 0.3 m deep bed of sand. Figure 1-14 illustrates the initiation of this filter's first production cycle.

A small increase in turbidity was observed over the first 15 minutes. The maximum turbidity recorded was 0.13 ntu, well within limits for filtered drinking water. Again, after 15 minutes, there was a decrease in turbidity until a constant level of 0.05 ntu was recorded.

From these results, it is evident that the measurement of turbidity with a continuous flow turbidimeter is both convenient and yields a smooth response that is readily interpreted.

Laboratory Measurements of Filter Effluent Turbidity

For confirmation, measurements of a sand filter's effluent turbidities, before and after backwashing, were made using a highly sensitive laboratory *ratio turbidimeter*. The results (Fig. 1-15) were very similar to those previously obtained with the continuous flow turbidimeters. Turbidity increased for 15 minutes and declined rapidly thereafter to levels within 0.1 ntu. This extended test sequence showed that turbidity continued to decrease over a period of 24 hours. This suggests that longer filter runs might minimize the suspended matter, as indicated by turbidity, entering the distribution system.

Effluent Turbidity from GAC-Capped Filter

A similar study was conducted with the effluent from a GAC-capped filter. Figure 1-16 shows that the first sample, taken five minutes after backwash, exhibited the highest turbidity (0.24 ntu). Thereafter, turbidity values declined to less than 0.1 ntu, as in previous studies.

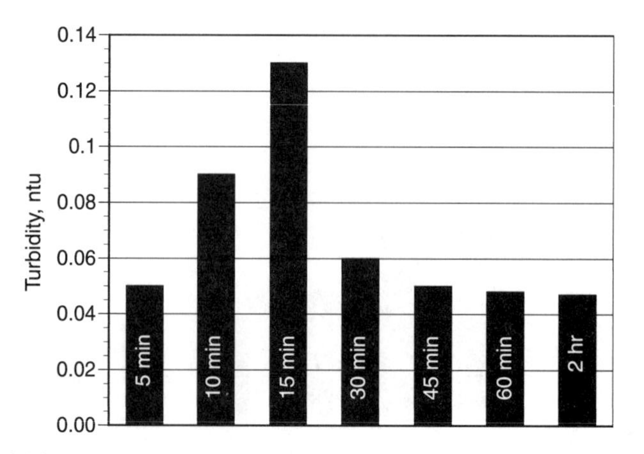

Figure 1-14 Turbidity of GAC-capped filter effluents after initial filter backwash.

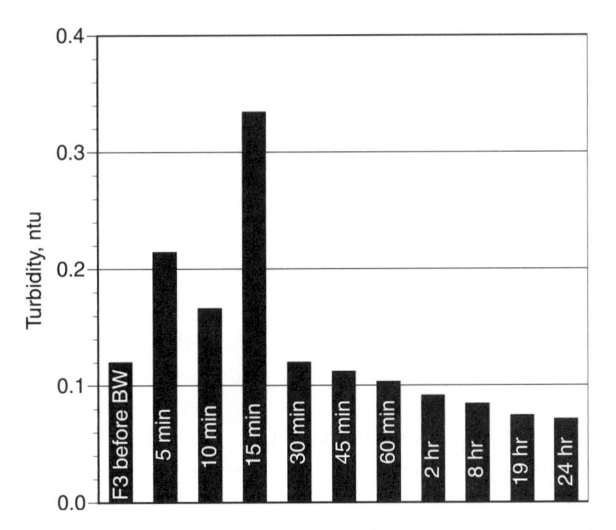

Figure 1-15 Turbidity of sand filter effluents before and after filter backwash.

Most of the turbidity in the Bloomington plant filtered water is caused by calcium carbonate and magnesium hydroxide floc. This is because large masses of these precipitates are formed as part of the lime softening (precipitation) process. From these results, it might be concluded that most of the solids that find their way into the Bloomington water distribution system were not initially in the source water, were formed during treatment, are not the result of anthropomorphic contamination, and do not represent a human health threat.

However, incorporated into the filtration process to remove organic solutes that occasionally contribute tastes and odors to the finished-water, GAC is another source of solid matter that may contribute particles to finished water. Compared with conventional filter sand, GAC is *friable*. It tends to abrade and release carbon fines to the finished water.

Carbon fines ranging in size from 0.5 to 2 μm can be visualized using the compound microscope. They can be differentiated from other particles by their black color and

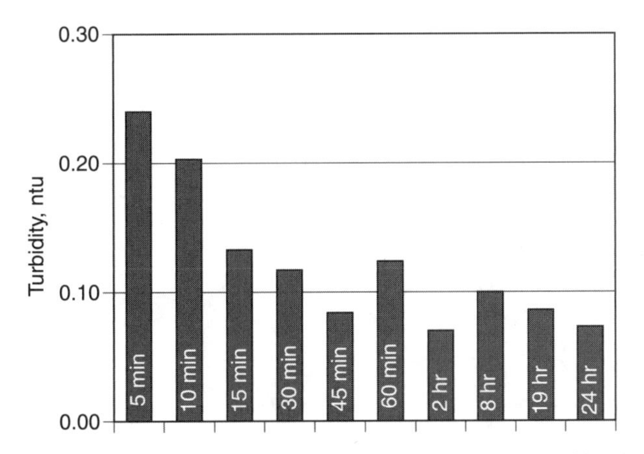

Figure 1-16 Turbidity of GAC-capped filter effluents after initial filter backwash.

asteroid-like appearance. The abundance of carbon fines in the filter effluent can be estimated after passing a measured volume through a 0.2 μm membrane filter.

ELECTRONIC PARTICLE COUNTING FOR EVALUATION OF FILTER PERFORMANCE

Electronic particle counters offer the potential for enumerating particles in various size ranges without separation by membrane filtration or size estimation using a microscope reticle. Their light-blocking sensors primarily detect dense, opaque, and dark-colored particles. Biotic particles, many of which are translucent, are detected poorly. Electronic particle counters may also either shear fragile aggregates of particles into fragments or record clumps of bacterial cells as single particles.

Figure 1-17 illustrates the numbers of particles in six size ranges detected by the Bloomington particle counter as a function of time after filter backwash. The first sample, taken before backwashing, represents the filter effluent at the end of a normal filter cycle. Forty particles per milliliter (24% of the total) were detected in the 2 to 3 μm size range while a maximum of 47 were detected in the 3 to 5 μm size range. While the particle counter reported detecting 20 particles per milliliter larger than 15 μm in the filtered water, such large particles were not generally evident in microscopic observations.

The effluent sample taken 15 minutes following backwash exhibited the highest total number of particles detected during the study, 381 per milliliter. Of these, 237 (62%) fell in the 2 to 3 μm particle size range, whereas an average of only 0.6 particles per milliliter exceeded 15 μm. During this study, the electronic particle count data consistently made it appear that larger particles were near-absent whenever smaller particles were abundant, an inexplicable inverse relationship. Only the combined electronic particle count data for the 2 to 3 and 3 to 5 μm size ranges appeared to reflect the pattern observed by turbidity monitoring.

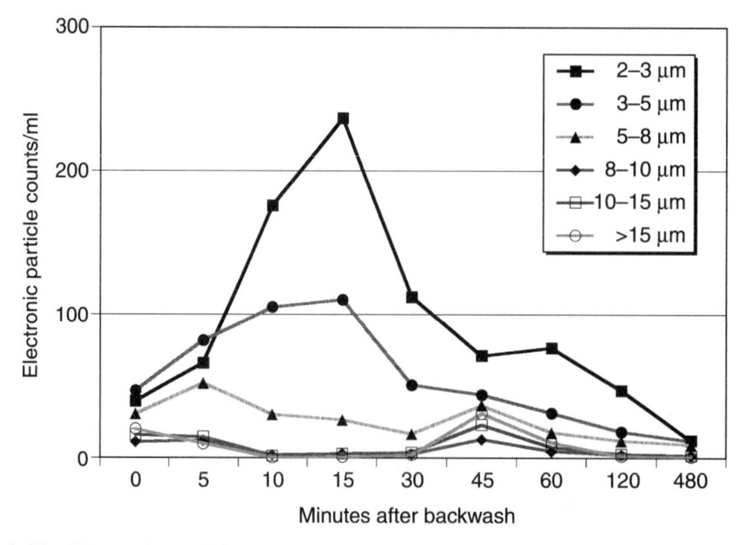

Figure 1-17 Electronic particle counts in sand filter effluent before and after filter backwash.

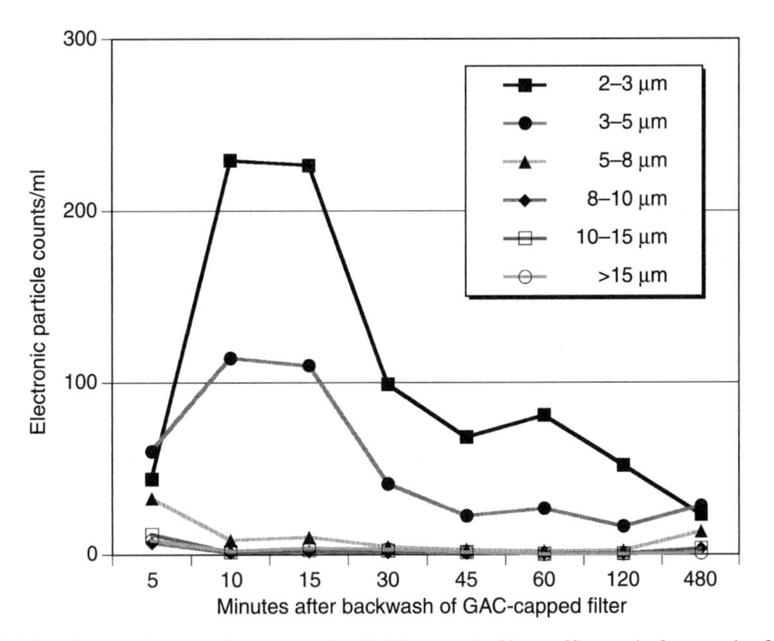

Figure 1-18 Electronic particle counts in GAC-capped filter effluent before and after filter backwash.

Electronic Particle Count Data Following Backwash of the GAC-Capped Filter

The results of a particle count study using another GAC-capped filter effluent (Fig. 1-18) were similar to those obtained in the previous study. Again, particles in the 2 to 3 and 3 to 5 μm size ranges exhibited peaks following backwash. Particles larger than 5 μm were nearly absent.

Since there was no obvious difference in the numbers of smaller (2 to 5 μm) particles detected in the studies of the two filters, it was inferred that carbon fines were either not present in significant numbers or not detected.

Overall, the total numbers of particles per milliliter detected by the electronic particle counter were extremely small (40 to 400/ml) compared with the number (>500,000/ml) that could be visualized using microscopy. The electronic particle counter appears to enumerate only that small fraction of particles that are opaque to the passage of infrared laser light. In addition, from the inverse relation between small and large particles detected, there is reason to suspect that larger particles may have been ruptured and dispersed during passage through the capillary *view volume* of the detector.

PHOTOMICROGRAPHS USING EPIFLUORESCENCE MICROSCOPY

A major advantage in using microscopy for particle analysis is its ability to enable visualization and identification of the wide range of particles present. Special, as well as commonplace, particles can readily be photographed for archiving and referral.

Algae and diatoms are common constituents of lake waters. The diversity, number, and dominant species of algae vary with season, temperature, rainfall, runoff, nutrient inflow, and numerous other factors. Recently, concern has emerged over the presence of blue–green algae because of their ability to produce microtoxins. Utility laboratory personnel should find it useful to learn to recognize these particular algae.

LIGHT MICROSCOPY

Bloomington's microscope system is capable of displaying, sizing, and enumerating various particle types. In the present study, ultraviolet light was used for epifluorescence microscopy. This technique allows the counting of particles containing RNA and DNA. The fluorescing particles are, primarily, bacteria and algae, as well as amorphous organic debris. The morphology (shape) of the organisms can be described and photographed for archival purposes. Most bacterial cells are small, short rods, whereas some reside in filaments or are attached to debris, such as decaying algae or insect parts. In many instances, a variety of bacterial cells seems to be embedded in organic film that may have detached from a solid surface. Depending on the DNA and RNA, as well as the activity of the cell, the color of fluorescence varies. Using acridine orange as a fluorochrome, the cells commonly fluoresce green, orange, or yellow.

Under a tungsten (white) light, microorganisms are not readily visible, but inorganic particles come into view. These may exhibit a wide range of apparent color and irregular shapes. In the case of calcium carbonate, the shape is crystalline, often exhibiting well-defined cubes. As noted, carbon fines are dark and may resemble asteroids.

The UV and tungsten lamp illumination can be used alternately or in conjunction. Newly formulated fluorochromes for UV microscopy can dramatically improve the intensity and visibility of fluorescing particles. Finally, the refracted light (Nomarski differential interference contrast) feature can provide a three-dimensional effect that other microscopic techniques cannot. In the present study, only the most fundamental total bacterial cell counting procedure was utilized to obtain data on organism populations in the filter effluents.

Water utilities make significant investments in particle monitoring equipment. Measuring devices have traditionally been based on light scatter (turbidimeters) and, in recent decades, the electronic particle counter. However, recent advances in membrane filters and organism staining procedures have now returned the microscope to its preeminent (referee) position for providing scientific insight to allow for better operation and, ultimately, optimization of physical and chemical water treatment processes for removal of particles of potential health significance.

Acquisition of a durable, fundamental tool, such as a basic microscope system, is consistent with the scientific management of a modern utility. Epifluorescence microscopy has emerged as a major tool for evaluating and improving water treatment processes. How valuable it will be depends largely on the interest, initiative, and creativity of the utility laboratory staff in using the microscope for visually and quantitatively assessing the impact of various treatment alternatives.

For many water utilities, the most prominent use of bacterial cell enumeration by microscopy to date has been in observing the recovery and growth of microorganisms in the distribution system. Biologically mediated corrosion and water quality deterioration during distribution have long been recognized as major causes of consumer complaints.

Appendix A provides guidance on procedures for total bacterial cell count by epifluorescence microscopy. Appendix B offers suggestions for potential studies involving microscopic particle analysis.

REFERENCES

APHA (1905). Standard Methods of Water Analysis: physical, chemical, microscopic and bacteriological methods of water examination (First Edition), American Public Health Association.

ASTM (1985). Enumeration of Aquatic Bacteria by Epifluorescence Microscopy Counting Procedure, D4455–85.

ASTM (1985). Simultaneous Enumeration of Total and Respiring Bacteria in Aquatic Systems by Microscopy, D4454–85.

ASTM (1988). Rapid Enumeration of Bacteria in Electronics–Grade Purified Water Systems by Direct Count Epifluorescence Microscopy, F1095–88.

Baylis, J. R. (1922). Microorganisms in the Baltimore Water Supply. *J. AWWA*, 9: 712.

Collins, V. G. and Kipling, C. (1957). The Enumeration of Waterborne Bacteria by a New Direct Count Method. *J. Appl. Bact.*, 20: 257–269.

Daley, R. J. (1979). Direct Epifluorescence Enumeration of Native Aquatic Bacteria: Uses, Limitations and Comparative Accuracy, Native Aquatic Bacteria: Enumeration, Activity and Ecology, ASTM STP 695, J. W. Costerton and R. R. Colwell, Eds., *ASTM*, 29–45.

Frankland, P. (1894). Micro-Organisms in Water—Their Significance, Identification and Removal. Longsmans, Green and Co., London.

Hobbie, J. E. et al. (1977). Use of Nuclepore Filters for Counting Bacteria by Fluorescence Microscopy. *Appl. Environ. Micro.*, 33: 1225–1228.

Kirchhoff, B. et al. (1988). Comparative Reductions of Turbidity, Heterotrophic Plate Count and Total Bacteria in a Ground Water, Lime Softening Plant. *Proc., AWWA Water Quality Tech. Conf.*, St. Louis, MO, 870–889.

Mittelman, M. W. et al. (1983). Epifluorescence Microscopy, A Rapid Method for Enumerating Viable and Nonviable Bacteria in Ultrapure–Water Systems. *Microcontam.*, 1: 32, 52.

Mittelman, M. W. et al. (1985). Rapid Enumeration of Bacteria in Purified Water Systems. *Med. Dev. and Diag. Ind.*, 7: 144.

Newell, S. Y. et al. (1986). Direct Microscopy of Natural Assemblages, Bacteria in Nature. Vol. 2: Methods and Special Applications in Bacterial Ecology, J. S. Poindexter and E. R. Leadbetter, Eds., Plenum Press, New York.

O'Connor, J. T. et al. (1982). Removal of Virus from Public Water Supplies, MERL, ORD, USEPA, Cincinnati, OH.

O'Connor, J. T. et al. (1985). Chemical and Microbiological Evaluations of Drinking Water Systems in Missouri, *Proc. AWWA Annual Conf.*, Washington, D.C.

O'Connor, J. T. (1990). An Assessment of the Use of Direct Microscopic Counts in Evaluating Drinking Water Treatment Processes. *ASTM Special Technical Publication 1102: Monitoring Water in the 1990's: Meeting New Challenges*.

O'Connor, J. T. and O'Connor, T. L. (2002). Control of Microorganisms in Drinking Water, Chapter 8: Rapid Sand Filtration. *American Society of Civil Engineers*, ISBN 0-7844-00635-9.

Palmer, C. M. and Tarzwell, C. M. (1955). Algae of Importance in Water Supplies. *Public Works*, 86: 107.

Pettipher, G. L. and Rodriques, U. M. (1982). Rapid Enumeration of Microorganisms in Foods by the Direct Epifluorescent Filter Technique. *Appl. Environ. Microbiol.*, 44: 809.

Pettipher, G. L. (1983). The Direct Epifluorescent Filter Technique for the Rapid Enumeration of Microorganisms. Research Studies Press Ltd, Letchworth, Herts. SG6 3B3, England.

Reach, C. D. et al. (1979). Virus and Bacterial Quality of Missouri River Water. *Proc. AWWA Annual Conf.*, San Francisco, CA.

Silvey, J. K. G. and Roach, A. W. (1964). Studies on Microbiotic Cycles in Surface Waters. *J. AWWA*, 56: 60.

Stevenson, L. H. (1978). A Case for Bacterial Dormancy in Aquatic Systems. *Microbial Ecology*, 4: 127.

Syrotynski, S. (1971). Microscopic Water Quality and Filtration Efficiency. *J. AWWA*, 63: 237–245.

Waksman, S. A. (1959). Actinomycetes; Nature, Occurrence and Activities, Williams and Wilkins Co., Baltimore, MD.

Whipple, G. C. (1899). Microscopy of Drinking Water, John Wiley, New York City.

2

PLANT PROCESS EVALUATIONS

Plant performance studies were conducted using Bloomington's newly acquired light microscope with epifluorescence attachment, video camera, and computer system to observe, quantify, and prepare micrographs of particles in suspension at each stage of treatment (Figs. 1-3a–1-8c; see color insert). The source water (Lake Evergreen during winter months) contained a diverse community of algae, bacteria, organic debris, and inorganic particles. Most of these particles were removed during lime softening which, in turn, produced particles of inorganic precipitates.

In an effort to improve overall plant particle removal efficiencies, additional assessments were made employing *flow ramping* to reduce solids penetration following filter *return-to-service* after backwash. The effectiveness of flow ramping was evaluated using three separate analytical measures: light-scattering (turbidity), electronic particle counting, and light microscopy.

PARTICLES OBSERVED AT VARIOUS WATER TREATMENT STAGES

Following Coagulation with Polymer and Ferric Sulfate

Cationic polymer is added at the Evergreen Lake intake so that the influent to the rapid mix receives water that is already partially coagulated. Thereafter, ferric sulfate, an inorganic coagulant, is injected at the rapid mix so that additional coagulation, plus a brown discoloration from the precipitation of ferric hydroxide, is observed as this water continues into the lime softener/clarifiers.

Water Treatment Plant Performance Evaluations and Operations. By John T. O'Connor, Tom O'Connor, and Rick Twait

Following Lime Softening

At the base of the softener/clarifiers, substantial quantities of slaked lime slurry are added to precipitate both calcium carbonate and magnesium hydroxide at pH reactions in excess of 11. Since most of the softening precipitation takes place within minutes, the solids produced are retained in the basin to form a dense *slurry blanket.* As the tangentially-injected influent water spirals upward through this blanket, it allows the lime-treated water to contact previously precipitated solids. This slurry blanket *solids contact* procedure is credited with allowing for more complete precipitation of hardness and increased growth of the particles of precipitated solids.

The effluent from the softener/clarifiers still contains any residual calcium carbonate and magnesium hydroxide that is not effectively removed by sedimentation. This freshly precipitated calcium carbonate, not initially present in the source water, may appear as small single cubic crystals or larger aggregations of small crystals. These comparatively clear particles are not readily stained by fluorochromes and may be best viewed under tungsten illumination since they do not fluoresce under ultraviolet light. However, at this stage of treatment, they are often the most abundant particles present in terms of both number and mass.

Following lime softening, the nature of the particles that are detected as turbidity or that can be enumerated by electronic particle counting has changed dramatically. The bulk of the particles found in lime-softened water were formed as precipitates during treatment. They were not in the source water to begin with. Efforts to estimate percent particle removals or plant efficiency using either turbidity or particle count data are intrinsically flawed if the plant process itself creates particles during treatment. Unlike the diverse particles found in the surface water sources, no health effects are associated with the particles created due to precipitated iron oxides, calcium carbonate, or magnesium hydroxide.

Additional benefits of lime softening include the co-precipitation of a wide range of trace metal oxides and carbonates, including lead, copper, cadmium, and zinc. Moreover, lime precipitation removes a small portion of the dissolved organic carbon present in surface waters. Finally, at high pH, sensitive microorganisms, such as viruses, may be inactivated.

Recarbonation following softening and settling results in the lowering of pH and the partial dissolution of some of the calcium carbonate crystals formed, thereby reducing turbidity. However, if an excess of lime (*overfeed*) has been applied during softening, recarbonation will result in the formation of additional calcium carbonate, thereby visibly increasing turbidity in the recarbonation basin. This lime overfeed may result in shorter filter runs and, in some instances, filter turbidity breakthroughs. To avoid these adverse effects, operators observe and sample the effluent from the recarbonation basins and measure its turbidity every two hours.

Following Filtration

Since filtration results in the removal of most of the calcium carbonate, the particles remaining in the filter effluent appear to be either *planktonic bacterial cells* that have passed directly through the filter or *attached bacterial growth* that has sloughed from the filter media.

As the Bloomington filters have been *capped* with granular activated carbon (a reducing agent that consumes oxidizing agents, such as chlorine), the practice of chlorination before filtration has been discontinued. Since a portion of the dissolved organic matter in the filter influent is adsorbed to the GAC, it can then serve as a substrate for microbial growth. While much of this attached growth is removed during each backwash, a fraction appears to become hydraulically stripped from the filter media during the normal filtration cycle.

With the influent algal and bacterial populations markedly reduced during lime softening and the calcium carbonate removed during sedimentation and filtration, many of the particles observed in the filter effluents would appear to have formed in the filter itself. Because of this, it seems unlikely that their presence is an indication of the penetration of potentially pathogenic organisms originating from the source water.

Following Disinfection and Clearwell Storage

The final barrier to the intrusion of pathogenic organisms at Bloomington is provided by chemical disinfection (chloramination). As required by federal and state regulations, Bloomington maintains a disinfectant residual concentration (C) for a prescribed time (t) such that a 90% destruction of any residual pathogen population is achieved. Combined, physical removal plus disinfection is intended to achieve a 99.9% reduction in the most resistant pathogens.

Micrographs (Figs. 1-8a–1-8c; see color insert) taken at the entry and outlet of Bloomington's 7600 m^3 finished water storage reservoir (*clear well*) indicate that the bacterial cell populations are low, but numerically similar. While differences in the color intensity of the fluorescence of the remaining cells may serve to indicate that most of these remaining cells have been *inactivated* by the added disinfectant during storage, the significance of the shift in color from robust green (presumably, active) to faint orange (presumably, inactive) is still a subject of scientific debate and investigation. While many of these cells may have been inactivated and cannot replicate, they have not yet lysed.

The present microscopic survey shows that particles in the source water have become aggregated by organic and inorganic coagulants. Their growth to form larger particles facilitates their removal by sedimentation. Still other source water particles have been entrained in the copious accumulations of inorganic precipitates formed by lime softening. This entrainment provides an additional degree of particle, including organism, removal that does not occur in water treatment plants employing coagulation alone.

For more widespread viewing by plant personnel, the particles observed microscopically at each stage of treatment can simultaneously be displayed on a computer monitor through a video camera and video capture board. From these displays, representative images can be stored on the computer's hard drive. Finally, the images may be transferred to a disc for archiving and viewing on other computers.

The micrographs not only illustrate the morphology of the particles found at each stage of treatment, but allow estimates to be made of their number and size. While it is not yet well known whether the apparent colors of the images provide further insight into the stage of development or activity of the microorganisms present, newly-formulated fluorescent stains promise the potential for distinguishing between live and dead cells. Such a distinction would be a valuable means for assessing the progress of disinfection, as well as for observing organism recovery and regrowth during distribution.

Electronic Particle Counts from Source to Filtered Water

For comparison with microscopic enumeration, electronic particle count data were obtained. Lake Evergreen water was found to contain a wide range of particle sizes, but relatively few exceeding 15 μm. At 13,000 counts per milliliter, the lake water contained about 15 times the number of particles detected in the finished water.

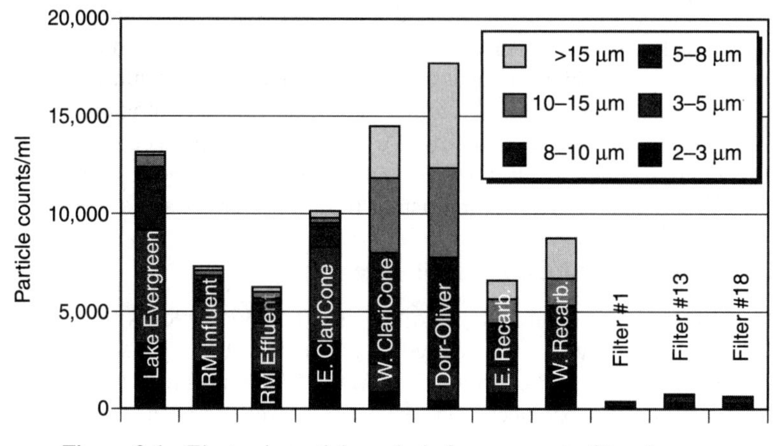

Figure 2-1 Electronic particle analysis from source to filtered water.

Particle count numbers appeared to decrease following the addition of coagulant, likely owing to the aggregation of numerous smaller particles into fewer larger ones.

As expected, the addition of large quantities of lime for softening resulted in the detection of more and larger particles in the effluents from the three softener/clarifiers. A large proportion of these particles exceeded 15 µm.

The recarbonation basin provided substantial additional particle removal through sedimentation and particle dissolution. Particles larger than 15 µm remained abundant in the recarbonation basin effluents/filter influents while comparatively few particles larger than 8 µm were detected in the filter effluents.

Figure 2-1 compresses the results of electronic particle counting data into stacked bar graphs. The bars illustrate the increases in the numbers and sizes of the particles observed following softening. Recarbonation (lowered pH) then reduces the numbers of total particles, but the larger particle sizes are seen to be more abundant than in the plant influent. Finally, the filtered waters were virtually free of large particles. The overall removal of particles, as indicated by the electronic particle counter, is somewhat misleading in that the sensor does not count smaller particles with the same efficiency as larger particles.

IMPROVING FILTER PERFORMANCE FOLLOWING BACKWASH

Filter Ramping Studies

Additional studies were undertaken to observe the benefit of ramping on filter performance. Ramping slowly brings the hydraulic loading on filters back to full service. This is done to minimize the passage into the filter effluent of particles dislodged from the filter media during backwash.

A dual media (GAC/sand) filter in the older filter gallery (Annex) was the first filter studied. All 12 filters in this gallery undergo ramping but each is also influenced by the operation of the other filters in the gallery. Due to plant operational requirements, the filters in this gallery distribute their total hydraulic loading. When one filter is taken out of service, the others share the total hydraulic load as the water level rises equally above all operational filters.

The performance of the test filter was evaluated by measurement of turbidity, electronic particle counts in various size ranges, and total bacterial cell counts by epifluorescence microscopy. Filter effluent samples were taken just before backwash and at successive times following backwash. Sampling was most frequent during the first two hours after the filter was returned to service. This was the period when the effects of backwashing on the number of particles passing the filter were most evident. Computer-controlled ramping in the gallery then took place over the first 30 minutes of filter operation.

Ramping is controlled by rate-of-flow-controlling butterfly valves on the effluent sides of the filters. Because these valves open incrementally, flows may momentarily exceed the *set point* flow until the system adjusts to close the valve slightly. Thereafter, the flow may be lower than the set point flow until another correction is made. These successive corrections result in modest flow variations as the flow controller *hunts* to find the set point flow. Flow variations due to this phenomenon were noted in the operation of this filter. Operationally, it is possible to overcome such variations by *locking* the rate-of-flow controller into position for selected periods.

Turbidity is the most commonly used and convenient measure of filter particle removal performance. Dedicated, on-line turbidimeters monitor each filter effluent and the data is recorded continuously throughout each filter cycle. Periodically, grab samples are also checked using the calibrated laboratory turbidimeter to confirm the stability of the online turbidimeters.

The turbidity data from the filter study (Fig. 2-2) indicates that turbidities remained low immediately following backwash and throughout the 30 minute ramping period. Even for most of the first hour of filter operation, turbidities were less than 0.3 ntu.

Electronic particle count data (Fig. 2-3) also indicated little disturbance of filter performance due to backwash. The total counts of particles larger that 2 μm were fewer than 1000 per milliliter except for the sampling at 45 minutes. This was also the time when turbidity peaked at 0.33 ntu. However, electronic particle count data did indicate that some relatively large particles (>8 μm) continually passed into the filter effluent. Alternately, bacterial cell counts showed a well-defined peak following backwash (Fig. 2-2). Cell count may be the most sensitive indication of the detachment of attached cells and organic debris from filter media during backwash.

Overall, turbidity measurement indicated that the Filter 1 performed satisfactorily following backwash, producing an effluent that would consistently meet proposed turbidity standards throughout the entire filter cycle. Moreover, since a significant portion of the turbidity in each of Bloomington's filter effluents may result from the softening process

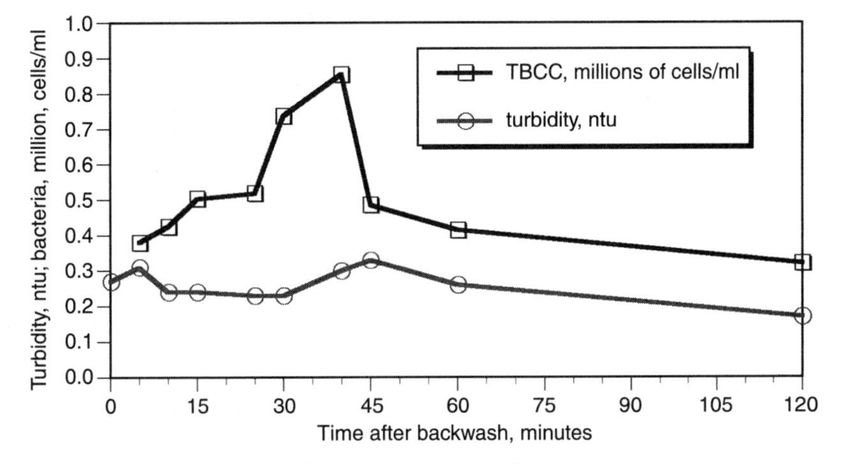

Figure 2-2 Effect of filter flow ramping on turbidity and total bacterial cell count.

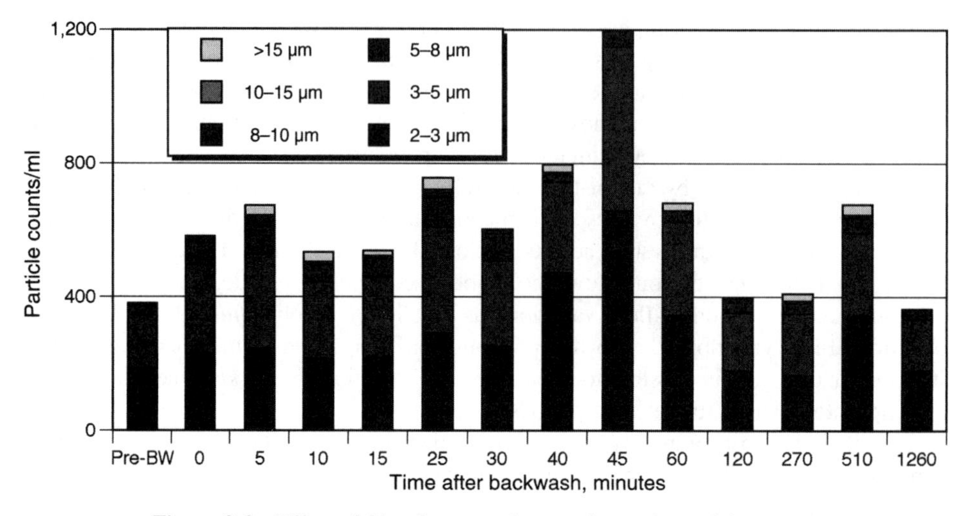

Figure 2-3 Effect of filter flow ramping on electronic particle counts.

producing calcium carbonate, it was proposed that the turbidities of Bloomington's filter effluents and finished water be evaluated after the water has been acidified. Following acidification, any calcium carbonate precipitate should redissolve and the turbidity be reduced solely to that caused by microorganisms, sloughed organic debris, and insoluble organic matter.

Correlation of Turbidity and Particle Count

A plot of filter effluent turbidity and particle count (Fig. 2-4) indicates a relationship between turbidity and particle count. A line of *best fit* intersects a turbidity of 0.3 ntu at a total particle count of 800 per milliliter. This linear correlation indicates a similarity of these two measurement techniques.

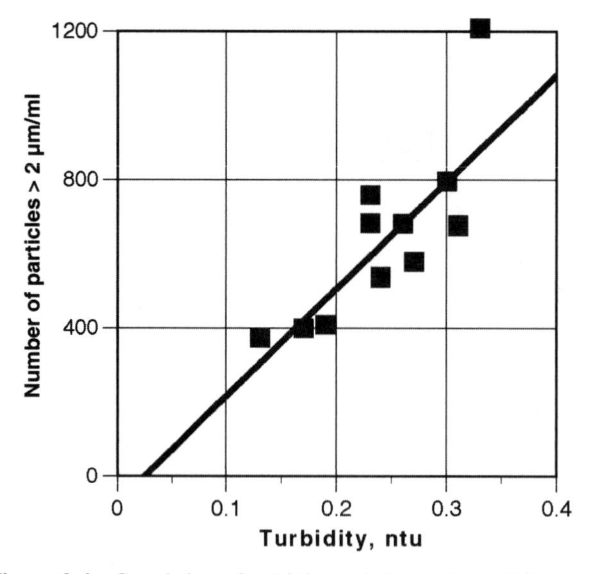

Figure 2-4 Correlation of turbidity and electronic particle count.

Filter Ramping Studies

A similar study was conducted using a filter in the new filter gallery of the plant operations building. In the initial study of this filter, ramping was not available. Modifications were in progress to allow computer-controlled ramping of the six filters in this gallery. Unlike the filters in the older gallery, these newer filters operate independently of each other.

In the absence of flow ramping, a turbidity excursion was observed. Turbidity peaked at 0.61 ntu 15 minutes after backwash, then declined over a period of approximately one hour to the levels observed before backwash. The initial peak and gradual decline in filter effluent turbidity was reflected in the electronic particle counts, which peaked after 10 minutes (Fig. 2-5). Counts of particles larger than 2 μm reached 5800 per milliliter at 10 minutes and declined slowly over the next hour.

Total bacterial cell counts were low before and immediately following backwash (175,000 cells/ml). After 10 minutes, cell numbers had increased to 460,000 cells/ml. Microscopic observation also showed that many calcium carbonate crystals were present. Occasionally, algal cells were observed. It was speculated that these algal cells might have been recruited from attached growth on the walls of the filter boxes.

The particle count results (Fig. 2-6) of a study of the second filter appeared to parallel the results of the initial study. Turbidity peaked 15 minutes following backwash and declined progressively over an hour. Electronic particle counts were also highest at 15 minutes and declined thereafter. Bacterial cell counts increased from 182,000/ml before backwash to 394,000/ml after 15 minutes.

While all three measurements reflected the impact of backwash on filter effluent, each appeared to be sensitive to different component parts of the spectrum of particles present in the filtered water. Turbidity is most strongly influenced by <1 μm (in the range of the wavelength of visible light) opaque, inorganic particles that absorb or scatter light most effectively. These might include calcium carbonate and inorganic precipitates as well as silt and clay particles.

The electronic particle counter is most sensitive to larger particles and less responsive to the far more numerous micrometer-sized particles. Since these larger particles potentially

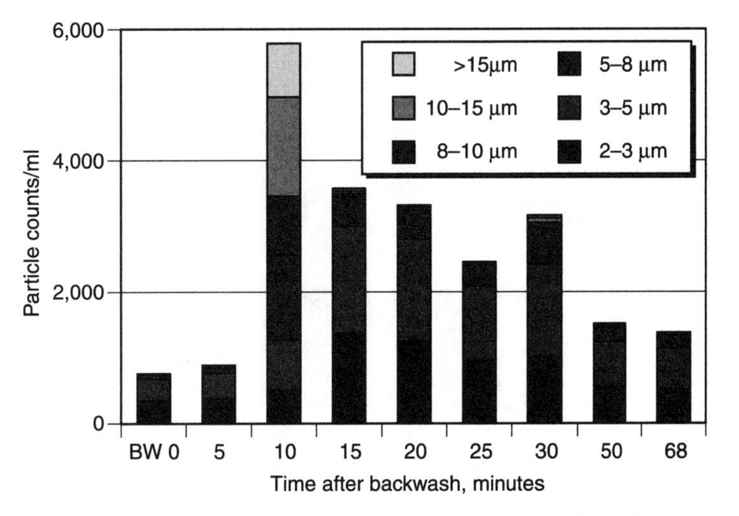

Figure 2-5 Effect of filter backwash (unramped) on electronic particle counts.

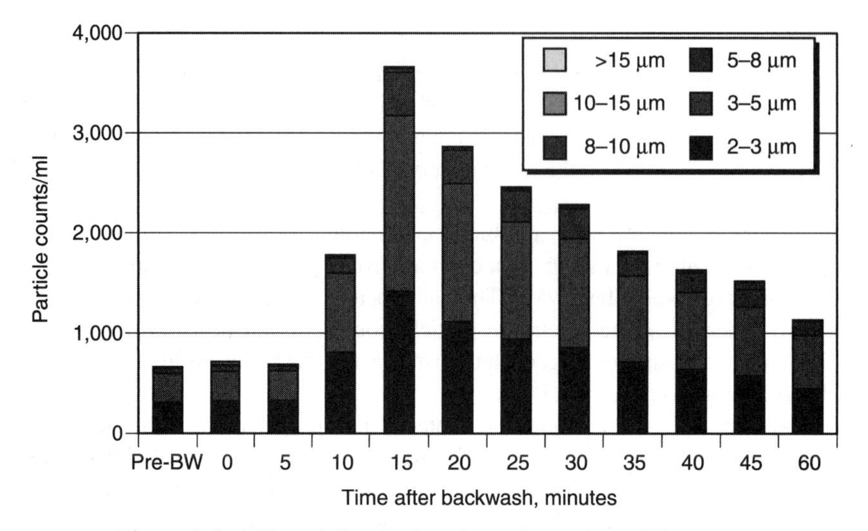

Figure 2-6 Effect of filter backwash on electronic particle counts.

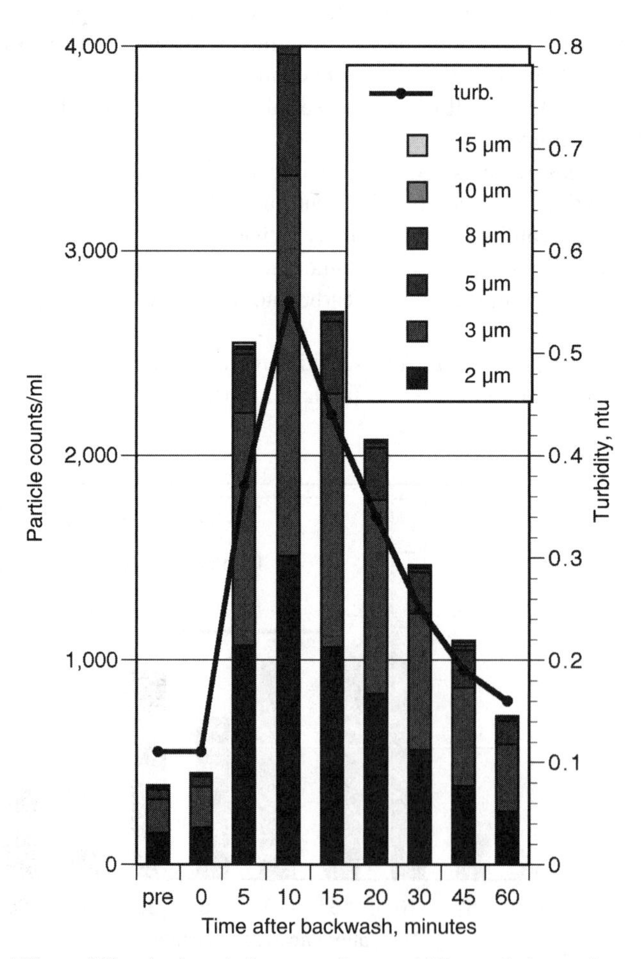

Figure 2-7 Effect of filter backwash (unramped) on turbidity and electronic particle counts.

provide microorganisms protection against subsequent chemical inactivation, they may represent particles of potential health concern.

Comparative Study of Ramping

Following the installation of flow ramping capabilities in the new filter gallery, a comparative study could be conducted. Initially, the test filter was not ramped. Turbidity and particle counts were measured for one hour following backwash (Fig. 2-7). Before backwash, the turbidity of the filter effluent was 0.11 ntu. After backwash, the turbidity remained at 0.11 ntu for the first sampling (0 minutes) before rising steeply to 0.55 ntu at 10 minutes. Thereafter, a gradual decline in turbidity to 0.16 ntu was observed after 60 minutes.

Particle count measurements reflected the turbidity excursion almost exactly. Less than 400 counts per milliliter were observed before backwash. This increased to over 4000 counts per milliliter after 10 minutes and declined thereafter to 700 after 60 minutes. The sum of particles counted in the eight samples taken following backwash was 28,109, whereas the average turbidity was 0.30 ntu.

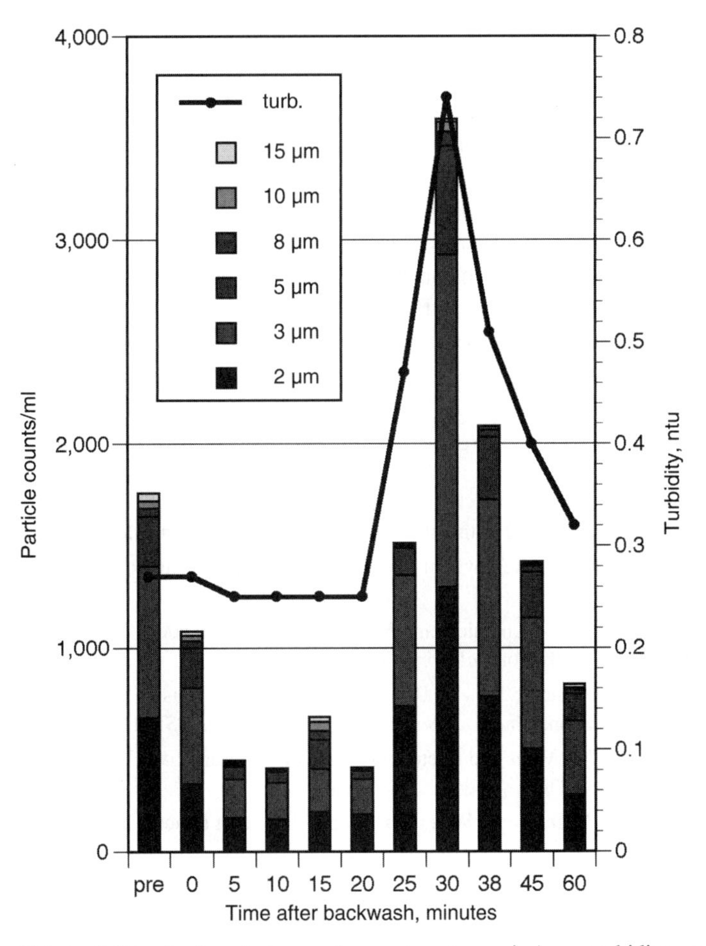

Figure 2-8 Effect of filter backwash (ramped on return-to-service) on turbidity and electronic particle counts.

Following a normal 48 hour filter cycle, the test filter was ramped for 30 minutes following backwash. Before backwash, the recorded turbidity registered 0.27 ntu (Fig. 2-8). Ramping for 30 minutes maintained this turbidity and delayed the turbidity excursion to the 30 minute sampling. Thereafter, turbidity peaked at 0.75 ntu. Simultaneously, particle count peaked at 3600 counts per milliliter.

Flow Ramping Study Results

The principal findings from these preliminary filter backwash studies indicated that:

- Ramping delays the onset of the turbidity excursion.
- Based on particle count data, the peak of the excursion is attenuated so that fewer total particles pass into the filter effluent. Most of the penetrating particles detected are smaller than 8 μm.
- Comparisons between filters may be inaccurate owing to differences in filter performance.
- There may be a performance advantage to a longer period of ramping, such as 60 minutes.

A comparison of ramping protocols following backwash for each individual filter should more precisely reveal the return-to service procedure that leads to the best filter performance.

The beneficial effects of ramping can be assessed by both turbidity measurements and particle counting. Since the results appear parallel, the online monitoring of turbidity remains the most convenient means of optimizing the backwash and flow ramping protocols for each individual filter. Bacterial cell counts may be utilized following backwash to aid in evaluating the effectiveness of the removal of attached bacterial cell mass from filter media.

REFERENCES

AWWA, ASCE (1998). Water Treatment Plant Design, Third Edition, McGraw-Hill.

O'Connor, J. T. et al. (1982). Removal of Virus from Public Water Supplies, MERL, ORD, USEPA, Cincinnati, OH.

O'Connor, J. T. et al. (1985). Chemical and Microbiological Evaluations of Drinking Water Systems in Missouri. *Proc. AWWA Annual Conf.*, Washington, D.C.

O'Connor, J. T. and O'Connor, T. L. (2002). Control of Microorganisms in Drinking Water, Chapter 8: Rapid Sand Filtration. *American Society of Civil Engineers*, ISBN 0-7844-00635-9.

Reach, C. D. et al. (1979). Virus and Bacterial Quality of Missouri River Water. *Proc. AWWA Annual Conf.*, San Francisco, CA.

Syrotynski, S. (1971). Microscopic Water Quality and Filtration Efficiency. *J. AWWA*, 63: 237–245.

3

LIME SOFTENING

PLANT PERFORMANCE EVALUATIONS

Treatment process evaluations were initiated in the winter of 2000–2001 to assess the softening performance of Bloomington's newest softener/clarifier (ClariCone) units. As part of the evaluation, data on softening and the removal of magnesium hydroxide and calcium carbonate were analyzed and compared with the results of jar test softening studies conducted over a range of pH.

In addition, the overall plant performance with respect to total organic carbon (TOC) reduction was assessed using newly acquired analytical equipment for the measurement of low concentrations of TOC.

Magnesium Ion Removals

Evaluations of plant data on softening unit performance indicated that magnesium removals varied seasonally. During periods of warm weather, magnesium hydroxide precipitated to levels approaching 10 g/m^3 (mg/l). Alternately, during winter months, magnesium concentrations in the plant finished water increased substantially, approaching 80 g/m^3. As a result, far more magnesium hydroxide was precipitated from the lake waters in the summer than in the winter. Since magnesium hydroxide is an effective coagulant, sludge blankets in the softening units are more stable when water temperatures are higher and precipitation of magnesium is more complete. Supplemental coagulants, such as polymers, are required, principally, when water temperatures are low. For the year 2000, magnesium in the Bloomington influent water averaged approximately 85 g/m^3 as calcium carbonate equivalent. Following lime softening, the delivered water averaged 39 g $CaCO_3$ eq/m^3. As shown in Table 3-1, magnesium removals averaged 54% while calcium removals averaged 29%.

Water Treatment Plant Performance Evaluations and Operations. By John T. O'Connor, Tom O'Connor, and Rick Twait

TABLE 3-1 Removals of Calcium and Magnesium by Lime Softening

Magnesium (2000 Avg.)	g $CaCO_3$ eq/m^3	g/m^3 as Mg
Influent (lake)	85	20
Finished	39	9.4
Removal	48 (54%)	10.6
Calcium (2000 Avg.)	**g $CaCO_3$ eq/m^3**	**g/m^3 as Ca**
Influent (lake)	111	44.4
Finished	79	31.6
Removal	32 (29%)	12.8

From this data, it appears that, on the average, 10.6 g/m^3 as Mg precipitated to serve as a coagulating agent for the influent suspended solids and precipitated hardness. This quantity of magnesium hydroxide precipitate, in itself, may have served as a primary inorganic coagulant and aided significantly in forming a stable sludge blanket in Bloomington's softener/clarifiers.

However, seasonal variations in monthly average influent magnesium ion concentrations as well as influent water temperatures influenced both the solubility of the magnesium hydroxide precipitate and the amount of coagulant floc formed.

The data plotted in Fig. 3-1, derived from the monthly average influent and finished water magnesium ion concentrations for the year 2000, illustrates the gradual decline in influent magnesium ion concentrations as the year progressed. Intermittent dips in influent magnesium ion concentrations may have resulted from dilution of the lake waters by rainfall and runoff.

Magnesium removals are indicated by the monthly average finished water magnesium ion concentrations. Seasonally, magnesium removals varied by a factor of eight. This variation was reflected in the larger amount of magnesium precipitated as magnesium hydroxide.

From these results, it is apparent that as much as three times more magnesium hydroxide floc is formed during the summer months. This precipitation should lead to far better

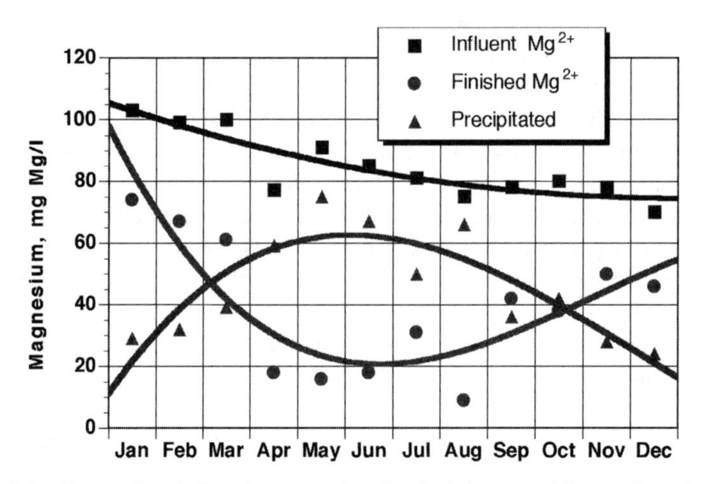

Figure 3-1 Seasonal variations in magnesium ion in influent and lime-softened effluent.

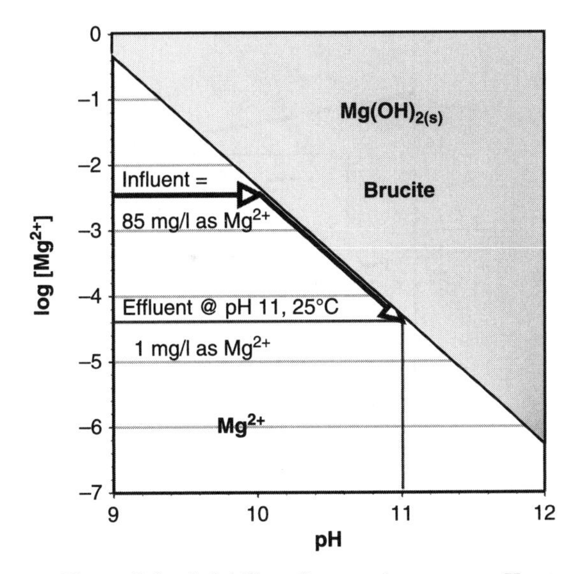

Figure 3-2 Solubility of magnesium versus pH.

flocculation and removal of the granular precipitates of calcium carbonate as well as the formation of a more stable slurry blanket.

An approximation of the solubility of magnesium ion in equilibrium with magnesium hydroxide (brucite) as a function of pH is illustrated in Fig. 3-2.

At the average lake water influent pH of 8.0, magnesium is highly soluble. Magnesium would be expected to remain in solution until pH exceeds 10 (horizontal black arrow). At higher pH, if equilibrium is attained, magnesium ion may be expected to precipitate as amorphous, flocculant magnesium hydroxide although the precipitation of the mineral, brucite, is assumed for purposes of estimating solubility. If such equilibrium is attained, magnesium ion would theoretically reduce to about 1 g/m^3 as Mg at pH 11 and 25°C.

$$pK_{sp} = 10.4 = pMg + 2pOH = 4.4 + 2 \times 3$$

The lowest average monthly concentration of magnesium observed in the Bloomington finished water was approximately 2 g/m^3 as Mg, achieved in August 2000. These results indicate that, under favorable conditions of temperature and pH, most of the magnesium ion in Bloomington's source water will be precipitated and recovered as magnesium hydroxide, $Mg(OH)_{2(s)}$, in the softening sludge. However, as influent water temperatures decline to 5°C, magnesium hydroxide solubility increases if pH is maintained at 11. This is partly because the ion product of water, K_w, decreases significantly with temperature.

$$\text{Ion Product, } K_w = [H^+][OH^-] \quad \text{and} \quad pK_w = pH + pOH$$

Figure 3-3 illustrates the effect of temperature on pK_w as well as its effect on the pH required to maintain a constant hydroxyl ion concentration of 1 mM/l (pOH = 3). Additional lime must be added to raise the pH from 11.0 at 25°C to approximately 11.7 at 5°C. If the pH is not increased at low temperature, the resulting reduction in hydroxyl

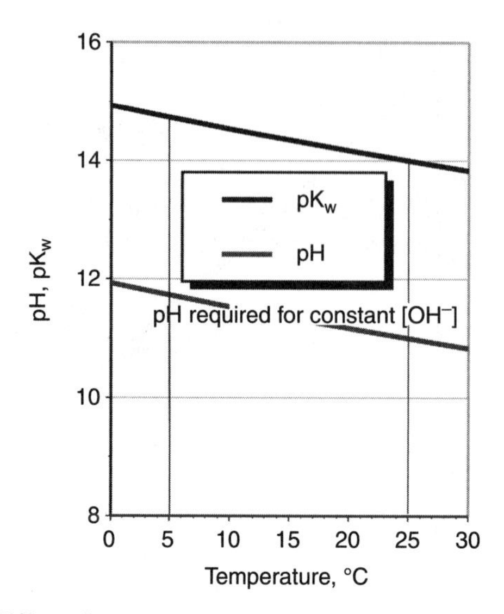

Figure 3-3 Effect of temperature on pH and magnesium solubility product.

ion concentration will increase magnesium hydroxide solubility. In addition, the kinetics of attaining solubility equilibrium is slower at lower temperatures.

Temperature also affects the solubility product, K_{sp}, for magnesium hydroxide (Fig. 3-4). As K_{sp} increases, $Mg(OH)_2$ solubility decreases by an order of magnitude over a temperature range of 0°C to 60°C. In other words, at constant pOH = 3, magnesium hydroxide is ten times more soluble at 0°C than 60°C. Not only do low temperatures retard the magnitude, but also the rate of $Mg(OH)_2$ precipitation.

$$K_{sp} = [Mg^{2+}][OH^-]^2 \quad \text{and} \quad pK_{sp} = pMg + 2pOH$$

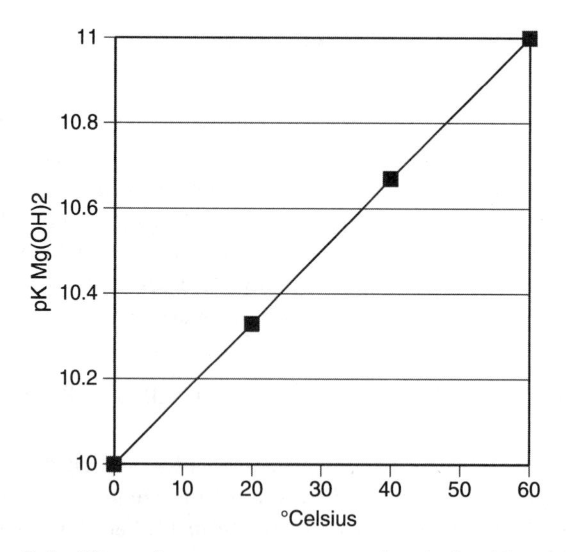

Figure 3-4 Effect of temperature on magnesium hydroxide solubility.

From an operational standpoint, the practical implication of these temperature-induced increases in magnesium hydroxide solubility is that it may be more difficult to maintain a stable slurry blanket in upflow softener/clarifiers unless alternative or supplemental coagulants, such as anionic polymers, are added when water temperatures are low. The increased addition of lime to increase pH and maintain pOH, the recovery and recycling of magnesium ion from the softening sludge, or an increased dosage of ferric coagulant may also help compensate for the increase in magnesium hydroxide solubility.

Seasonal increases in lime dosage may also necessitate the addition of larger quantities of carbon dioxide in order to maintain a constant finished water pH. While not indicated for Bloomington, some water softening processes utilize a secondary treatment stage of blending softened with unsoftened influent water to minimize this additional chemical requirement.

Hardness Reduction

The average annual reduction of hardness from the Bloomington influent to finished water is illustrated in Figs. 3-5 and 3-6. For the year 2000, lake water hardness was decreased from 195 to 111 g $CaCO_3$ eq/m^3, a 43% average reduction. As Table 3-2 indicates, Bloomington's finished water hardness falls within the range of values commonly found in natural and lime-softened midwestern waters.

The 80 g $CaCO_3$ eq/m^3 reduction in alkalinity is also consistent with the observed average reduction of hardness (84 g $CaCO_3$ eq/m^3). As expected, noncarbonate hardness is essentially the same in both the plant influent and finished waters.

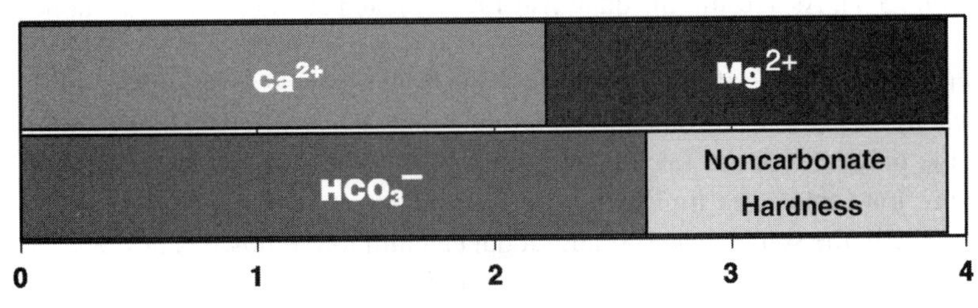

Figure 3-5 Electroneutrality condition for source water.

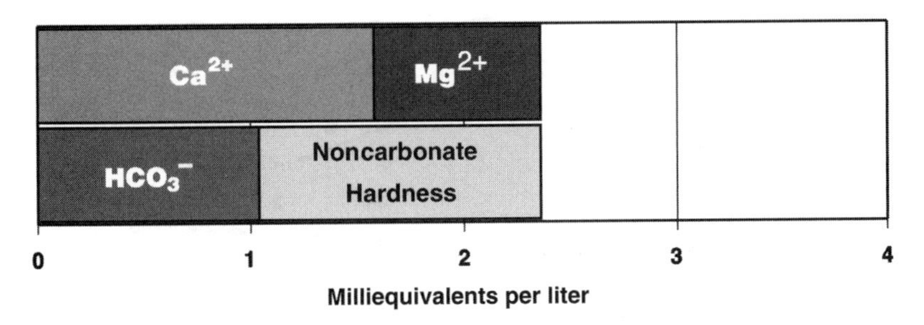

Figure 3-6 Electroneutrality condition for lime-softened water.

TABLE 3-2 Total Hardness in Source and Finished Waters, g $CaCO_3$ eq/m^3

City	Water Sources	Raw	Finished
Kansas City, MO	75% MO River; 25% alluvial wells	218	85
St. Louis, MO	66% MS River; 34% MO River	208	107
Columbia, MO	Alluvial wells in MO River flood plain	350	155
Chicago, IL	Lake Michigan; South District Plant	128	128
Highland, IL	Silver Lake	104	141
Normal, IL	14 Wells	419	108
Bloomington, IL	Lakes Bloomington, Evergreen	195	111

Lime Dosage

An estimated average of 145 g/m^3 of lime, CaO (129 g $CaCO_3$ eq/m^3) was added for softening in the year 2000. This dosage is approximately equal to the average Bloomington plant influent alkalinity (132 g $CaCO_3$ eq/m^3).

The minimum theoretical average lime dosage for excess lime softening (including magnesium removal) would be equal to the alkalinity plus the average magnesium carbonate concentration (21 g/m^3) for a total of 153 g $CaCO_3$ eq/m^3.

Calcium Removals

Bloomington influent calcium concentrations ranged from 97 to 141 g $CaCO_3$ eq/m^3, with highest values recorded in the spring. Similarly, during the year 2000, variations in effluent (finished water) calcium ion concentrations paralleled the monthly average influent concentrations (Fig. 3-7). As a result, the quantity of precipitated calcium carbonate appeared to be nearly constant throughout the year.

On the average, 29% of the influent calcium was removed in the softening process as compared to 54% of the influent magnesium. However, in September, approximately 50% of the influent calcium ion was removed.

Calcium ion removal is limited by the amount of alkalinity (bicarbonate + carbonate ion) available in the source water. If additional calcium ion removal is desired, the addition of sodium carbonate, Na_2CO_3, would be required.

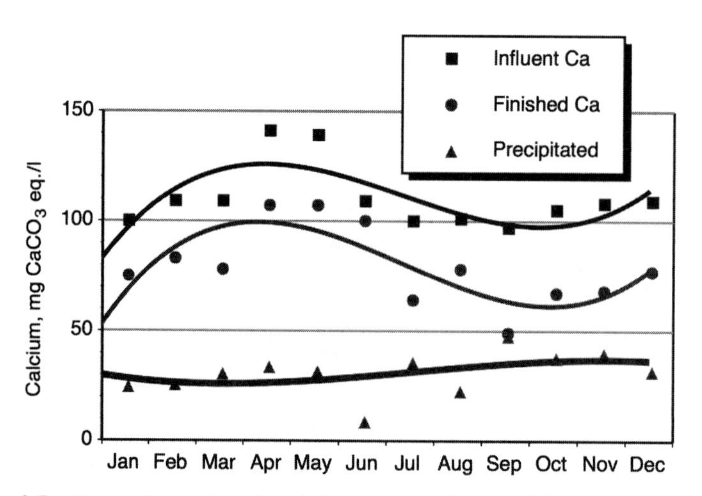

Figure 3-7 Seasonal variations in calcium ion in influent and lime-softened effluent.

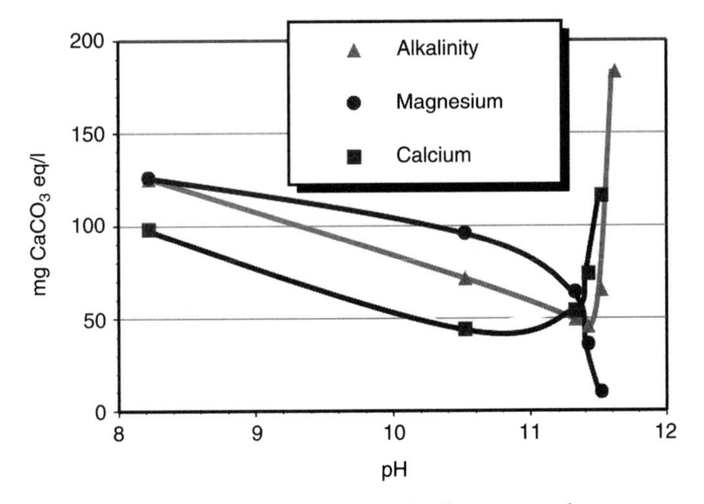

Figure 3-8 Effect of pH on hardness removal.

Jar Test Studies of Hardness Reduction

To observe the effect of varying pH on hardness removal, jar test studies were conducted on the plant influent water in January 2001. The results (Fig. 3-8) illustrate the progressive decrease of magnesium ion with pH. At low temperature, pH 11.5 yielded a magnesium ion concentration of 10 g $CaCO_3$ eq/m^3.

Calcium ion concentrations declined from 98 to a minimum of 46 g $CaCO_3$ eq/m^3 at pH 10.5. The carbonate ion concentration is near its maximum at this pH. At pH reactions as high as 11.5, calcium ion concentrations again increase due to the solution of added lime (calcium hydroxide). In this study, a minimum settled water hardness was achieved at about pH 11.4.

The subsequent recarbonation of this water to approximately pH 9 is expected to return any suspended magnesium hydroxide and calcium carbonate (*carryover* in the clarifier supernatant) to solution.

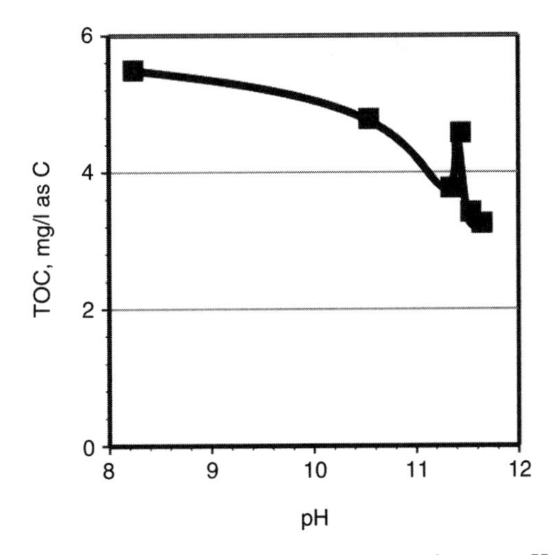

Figure 3-9 Total organic carbon removal versus pH.

These jar studies, coupled with new analytical instrumentation that allowed the measurement of both TOC and total inorganic carbon (TIC) provided an opportunity to assess the concurrent removal of both organic and inorganic carbon. TOC removals increased with pH and the quantity of precipitate formed. Figure 3-9 shows that over 40% of the TOC was removed at pH 11.6. This is closely equivalent to the TOC reduction observed at full-plant scale.

REMOVAL OF ORGANIC MATTER

As with the removal of microorganisms, the removal of organic matter occurred primarily in the softening units. Of the 43% TOC removal through the entire treatment process, 42% was achieved in the softener/clarifier units. These results indicated that, even during cold weather periods, the precipitative softening process achieved TOC removals far in excess of the 25% required by the U.S. Environmental Protection Agency (EPA) surface water treatment rule.

Table 3-3 confirms that most of the TOC removal took place during the softening process, probably as the result of coagulation and entrainment of the *particulate* organic matter. As viewed under the microscope, much of this organic matter appeared to be in the form of algal and bacterial cells, as well as organic debris, slime, and detritus.

Since little removal of organic matter was evident subsequent to softening, the *dissolved* portion of the TOC, sometimes referred to as DOC, was not reduced significantly under these winter operating conditions.

From these initial evaluations, it is evident that assessments of TOC removals should be made on a seasonal basis, particularly in response to lake water temperature changes. In addition, the transient effects of storm water influxes and induced mixing (destratification) on dissolved and particulate organic carbon should be quantified. However, since the observed TOC removal processes appear to be highly effective at low influent water temperatures, comparable or better removals would be expected as temperatures rise.

The monitoring of TOC removal allows the water plant operational staff to better manage the treatment process by making seasonal adjustments to chemical dosages, flows, filtration rates, and blending protocols that will further enhance the removal of a range of organic substances, including odorous compounds, herbicides, and the contributors to microbial growth during distribution.

TABLE 3-3 Removal of Total Organic Carbon during Treatment

Sampling Point	TOC, g C/m^3	Removal
Rapid mix basin	6.1	Influent
Softener (19,000 m^3/d)	3.54	42%
Softener (26,000 m^3/d)	3.64	–
Recarbonation—influent	3.48	–
Recarbonation—middle	3.56	–
Recarbonation—effluent	3.49	–
Filter 13—influent	3.46	–
Filter 13—effluent	3.46	43%

REFERENCE

Stumm, W. and Morgan, J. J. (1981). Aquatic Chemistry, Second Edition, Wiley-Interscience.

4

ACIDIFICATION PROTOCOL

REGULATION OF TURBIDITY AS A MICROBIOLOGICAL SURROGATE

The turbidity measurement derives primarily from the light-scattering and absorption properties of the particles suspended in water. Light scattering causes water to appear cloudy and, to a degree, colored. In natural waters, turbidity is imparted by silt, clay, metal precipitates, woody organic debris, hairs and fibers, pollen, algae, bacteria, plankton, oil emulsions, gas bubbles, and numerous other suspended particles. However, even when present in equal size and numbers, translucent particles, such as bacterial cells, contribute far less to measured turbidity than dense, opaque, sharply faceted particles.

The particles observed in natural water sources are extraordinarily varied. Some scatter light well, others are translucent. Some are dense and settle readily, others approach the density of water and remain suspended. While some particle surfaces may have a high charge density, others have virtually none.

In the case of microorganisms, surface properties and their tendency to attach to the surfaces of other, larger solids may vary with metabolic activity. The extent of organism attachment to particle surfaces even varies with seasonal temperature changes. However, the most obvious, observable differences in waterborne particles are their shapes (morphology). Particles may range from long, highly flexible rods and filaments to loose aggregations of diverse particles (microfloc) to rigid inorganic spheroids. In addition, biotic particles may be flexible and deformable, able to squeeze through filter pores and narrow passages. Many microorganisms are motile.

Owing to the potential for river and lake waters to become contaminated with waste discharges, the reduction of turbidity during surface water treatment has long been utilized, operationally, to assess the effectiveness of removal of agents of *potential health significance* (pathogenic or disease-causing organisms). For the past century, turbidity reduction

Water Treatment Plant Performance Evaluations and Operations. By John T. O'Connor, Tom O'Connor, and Rick Twait

Figure 4-1 Jar testing to determine optimum coagulant and lime dosages.

has been used, almost exclusively, for estimating the *microbiological efficiency* of the physical removal processes (sedimentation, filtration) employed in water treatment plants. Figure 1-3d illustrates the organisms accompanying and colonizing a decomposing *Daphnia.*

In drinking water treatment, chemical coagulants are generally added to aggregate the finer, micrometer-sized particles in the source water. If coagulation is successful, the resulting floc particles are large, denser then water, and readily settleable (Figs. 1-3l and 1-3m; see color insert).

As illustrated in Fig. 4-1, during flocculation and lime softening, the chemical additives themselves add substantially to the turbidity and solids content of the water. In this coagulant test series, a colored *control* sample shows the effect of adding iron coagulant, alone, to Lake Bloomington water. The other jars have been additionally treated with varying dosages of lime for softening. This illustration visually demonstrates that both coagulation and softening are particle-producing, as well as particle-removing, processes. However, the mass of precipitates produced by these chemical additions does not constitute particles of potential health significance.

ADVANCED ANALYTICAL METHODS FOR EVALUATION OF WATER TREATMENT PLANT PERFORMANCE

Finding it difficult to associate actual waterborne disease outbreaks with traditional measures of particle removal performance, the EPA has formulated ever more elaborate and restrictive regulations governing finished water turbidity. Attributing disease outbreaks to short-term exceedances of turbidity from individual filters, the EPA Surface Water Treatment Rule (SWTR), requires the continuous monitoring of each individual filter unit. Under the SWTR, each defined filter effluent turbidity excursion requires an individual report to the state regulatory (primacy) agency.

In the case of turbidity excursions at Bloomington, most of the particles found in filtered and finished water result from the massive and progressive formation of calcium carbonate and magnesium hydroxide during (and following) the lime softening process. The EPA now recognizes this as the result of *lime carryover* and has made special turbidity monitoring allowances for those water utilities practicing lime softening.

Calcium carbonate is commonly supersaturated in lime-softened waters and will tend to gradually approach equilibrium over long periods of time by *post-precipitation.* Post-precipitation during or even following filtration thereby creates a false assessment of a softening plant's ability to remove particles of potential health significance.

Upon application to their regulatory authority, U.S. utilities practicing softening by lime precipitation may be permitted to apply for approval of an *acidification protocol* in which filtered water samples are treated with acid to redissolve calcium carbonate prior to final turbidity measurement. Properly conducted, this modified procedure allows turbidity measurements to be more closely associated with those particles that were initially in the source water rather than those formed as by-products of the softening process. However, even with the removal of softening precipitates, the measured final acidified turbidity would likely underestimate particle removal efficiency from source waters owing to the recruitment of nonpathogenic particles during passage through the various water treatment units.

Another regulatory concession to utilities practicing lime softening is the allowance of applications for *alternate exceedance levels*. This authority enables state regulatory agencies to specify higher turbidity levels when it can be demonstrated that the higher turbidity levels observed are due to lime carryover and not treatment or filter failure.

In addition to the establishment of a scientifically defined acidification protocol, a major objective of the evaluation described here was to obtain plant operational data to determine the relative contribution of lime carryover to the turbidity observed in the Bloomington plant filter effluents. The acidification procedure demonstrated and adopted should redissolve the precipitates formed during the treatment process without eliminating any particles that might have originated in the source water. Specifically, treatment to dissolve these precipitates should not result in the *lysing* (bursting) of the microbial cells found in the source water.

Electronic Particle Counting and Sizing

Over the past decade, a number of water treatment plants have started to utilize electronic particle counters in an effort to quantify the particle removals achieved as a function of particle size. The Bloomington water department laboratory also employs an electronic particle counter as a supplement to turbidity and microscopic measurements to evaluate the performance of its particle-removing processes.

The special promise of the electronic particle counter (Fig. 4-2) is that it will permit enumeration of the particles in both source and finished water by size classification.

Figure 4-2 Particle enumeration and sizing using an electronic particle counter.

Since it is now known that the disease-causing cysts of *Giardia lamblia* and *Cryptosporidium* are, predominately, in the size range of 4 to 10 μm, the effectiveness of removal of particles in this specific size range, as opposed to overall turbidity reduction, may be estimated using particle count data.

As described in Chapter 2, Bloomington has also used its electronic particle counter to assess and automate its procedure for returning filters to service following backwash. This was done in conjunction with the installation of a complex, new filter *system control and data acquisition* (SCADA) system.

Epifluorescence Microscopy for Particle Identification

As described in Chapters 1 and 2, the Bloomington laboratory has been equipped with high-quality microscopic equipment so that the particles in the plant source and finished waters could be directly observed and enumerated on monitors, then identified and photographed.

Using fluorescent stains, the direct microscopic count allows the plant laboratory staff to rapidly observe, identify, and enumerate the major particles entering and penetrating the water treatment plant. The microscope can also be used to validate the implicit assumption that these particles or aggregates of particles can be electronically sized as to their *equivalent spheres*.

The combination of these three measurements allows the Bloomington plant laboratory staff to more critically observe and compare the removals of various groups of particles. In addition, they can observe seasonal variations in influent particle distributions (e.g., algal blooms) and treatment plant particle removal efficiencies as lake water temperatures vary from 0°C to 28°C. From these observations, chemical feed dosages can be more rationally optimized for difficult cold weather conditions or high influent turbidities.

FORMATION OF PARTICLES DURING TREATMENT

As shown in Chapter 2, lime softening plants serve as notable exceptions to plant performance evaluation schemes relying on turbidity or electronic particle counting. Bloomington's treatment process begins with the addition of an iron (ferric sulfate) coagulant plus a cationic polymer to the lake water during pretreatment. These agents increase the suspended solids while effectively agglomerating the numerous smaller particles in the lake water. Most of the smaller (micrometer-sized) particles entrained by coagulation are well below the size resolution of electronic particle counters. As a result of this primary coagulation, many larger, but fragile, aggregate particles are formed.

Coagulation is followed by the application of a large quantity of lime to precipitate substantial amounts of the hardness-producing ions, calcium and magnesium. The magnesium hydroxide formed is a highly effective supplementary coagulant and enables the formation of a *sweep floc* that aids in the formation of a stable slurry blanket in Bloomington's upflow contact clarifiers. Microscopic observations confirm the expectation that passage through this deep blanket of calcium carbonate and magnesium hydroxide provides extensive additional removal of those particles initially in the source water. Clarification (settling) above the slurry blanket then results in the reduction of turbidity to low levels.

However, subsequent recarbonation (the application of carbon dioxide gas for pH reduction and chemical stabilization) may result in further precipitation of additional calcium carbonate if the carbon dioxide reacts with excess residual lime (Fig. 4-3).

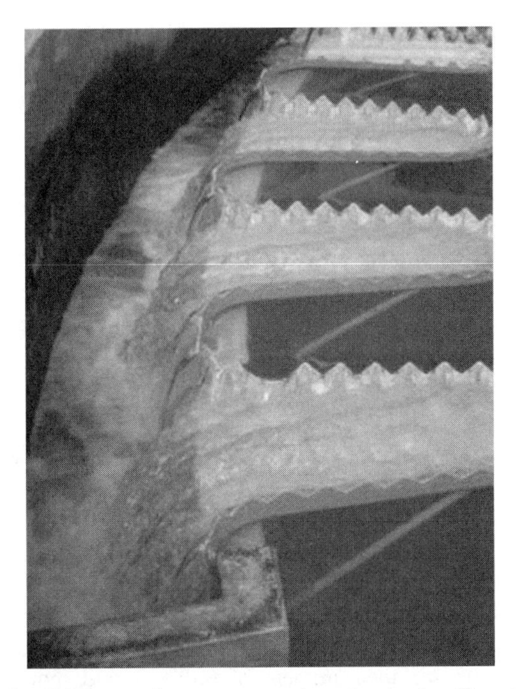

Figure 4-3 Calcium carbonate on recarbonation basin effluent launders.

Despite the implications of any observed increases in turbidity of the recarbonated water, few particles initially contained in the influent source water remain in suspension.

By any measure, it is evident that far more particles are formed during softening by precipitation than ever were in the source water. The mass of solids following the addition of lime to Lake Bloomington water may be 100 times greater than was initially in the water. The production of these innocuous solids leads to a serious underestimation

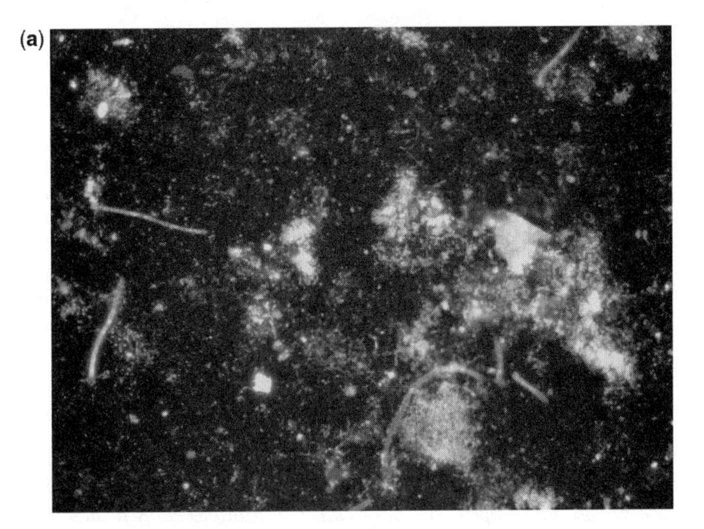

Figure 4-3a Material rinsed from filter media indicate abundance of attached growth of bacterial cells and filaments largely embedded in slime. (See color insert.)

of the efficiency of removal of the particles of potential health significance. In virtually all water treatment plants where particles are generated or modified during treatment, the inability of the turbidity measurement or the electronic particle counter to discriminate native from process-generated particles results in a gross underestimation of water treatment plant performance efficiency. In addition to masses of calcium, magnesium, iron, aluminum, and manganese precipitates, particles generated or recruited during treatment may include paint chips, rust particles, powdered activated carbon *fines*, and even, oil droplets. Where basins are open to sunlight, additional algal growth recruited from basin walls generally adds to the particles in suspension. Finally, the intermittent sloughing of fresh bacterial growth and surface accumulations from filter media (Fig. 4-3a; see color insert) may contribute to measured finished water turbidities.

IMPLICATIONS FOR ASSESSMENT OF WATER TREATMENT PLANT PROCESS EFFICIENCY

A willingness to observe and characterize the particles present in natural waters, with a special emphasis on the biotic particles of potential health significance, is a prime prerequisite to the establishment of a rational means for evaluating treatment plant performance. Accordingly, the current study was aimed at developing an analytical protocol that would control interferences with the measurement of turbidity and more accurately assess the degree of reduction of source water turbidity. This acidification protocol focuses on returning the dominant particle created during the softening process, calcium carbonate, to solution. The residual turbidity would, then, more closely reflect the degree of removal of particles that might have been, initially, in the source water.

Partly based on the results of this study, regulatory agencies may allow utilities to submit a scientifically defined acidification protocol for approval. Moreover, based on the operational results obtained following acidification, utilities may apply for alternate exceedance levels that more accurately reflect the turbidity of the finished water when lime carryover is accounted for.

INTERIM ENHANCED SURFACE WATER TREATMENT RULE

The following excerpts from the final IESWTR address the measurement of turbidity at lime softening plants and provide for alternative exceedance levels in order to limit the continuing need for the acidification of individual samples.

> *40 CFR Parts 9, 141, and 142: National Primary Drinking Water Regulations: Interim Enhanced Surface Water Treatment; Final Rule: Sec. 141.173 Filtration.*
>
> . . .
>
> *(3) A system that uses lime softening may acidify representative samples before analysis using a protocol approved by the State.*
>
> *Systems that use lime softening may apply to the State for alternative exceedance levels for the levels specified in paragraphs (b)(1) through (4) of this section if they can demonstrate that higher turbidity levels in individual filters are due to lime carryover only and not due to degraded filter performance.*

ACIDIFICATION

Turbidity as a Function of pH

In initial tests, four-liter grab samples from the Bloomington Water Treatment Plant were progressively acidified in the laboratory with a strong acid (1 : 1 HCl). Turbidity and pH were measured. Select subsamples at varying pH were observed microscopically and/or passed through an electronic particle counter.

A lime-softened, settled water sample was taken from the effluent of the recarbonation basin, before filtration. The initial turbidity was relatively low at 0.5 ntu. The pH versus turbidity curve acquired from the stepwise addition of HCl acid is shown in Fig. 4-4. Turbidity decreased to 0.2 ntu with the addition of acid to pH 2. In this instance, the dissolution of calcium carbonate resulted in a 60% reduction in turbidity.

A second sample was a mixture of clarified water and slurry blanket solids drawn from depth within the softener, before recarbonation. The initial turbidity of this mixture was intentionally selected to be relatively high at 15 ntu.

The pH versus turbidity curve for acidification of this sample, shown in Fig. 4-5, again shows that turbidity progressively decreases with pH, reaching a low of 2.0 ntu at a pH of 2. This reduction represents an 87% decrease in turbidity upon acidification.

Both tests showed significant decreases in turbidity until a pH of 4 was reached. Beyond that point, there appeared to be decreasing additional turbidity reduction associated with acidification to pH 2.

Particle Counts

An electronic particle counter was used to quantify the number of particles in subsamples from the acidification of the clarifier effluent. Figure 4-6 shows the breakdown by particle size range of the number of particles counted at various pH values.

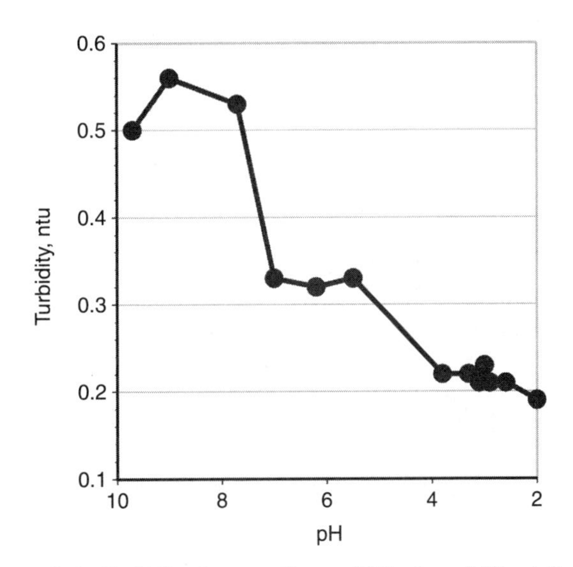

Figure 4-4 Turbidity decrease from acidification of filter influent.

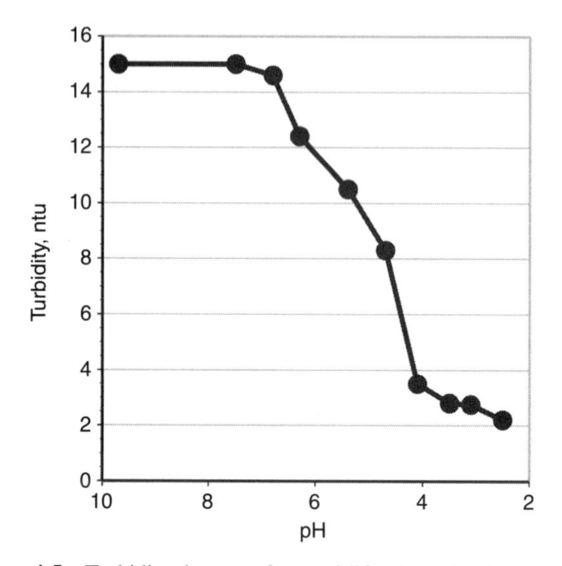

Figure 4-5 Turbidity decrease from acidification of softener solids.

Total particle numbers were lowest at pH 2.5, the lowest pH tested. The greatest percentage decrease (92%) was in the number of particles 15 μm and larger.

Figure 4-7 is similar to Fig. 4-6. It has been modified to illustrate the difference in the effect of acidification on large versus small particles. The number of particles >8 μm was reduced by approximately 85%, while those <8 μm was reduced by only 44%. This is because the calcium carbonate crystals in this slurry blanket sample were relatively large compared to silt, microorganisms, and organic debris.

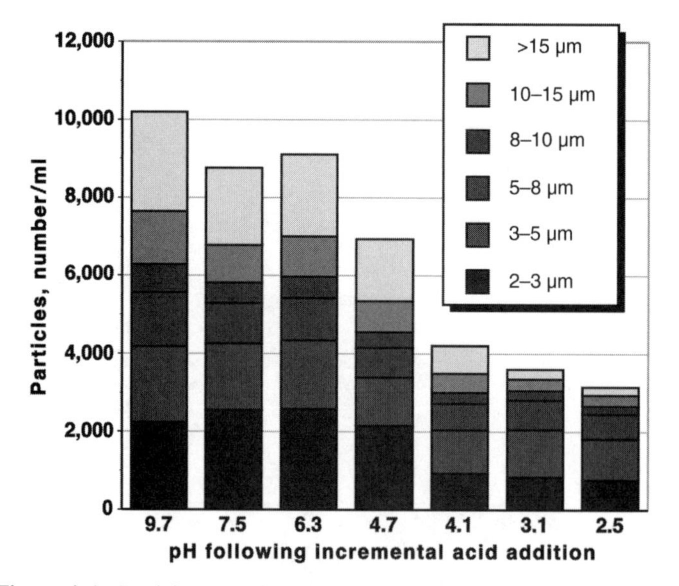

Figure 4-6 Particle count decreases with acidification of softener solids.

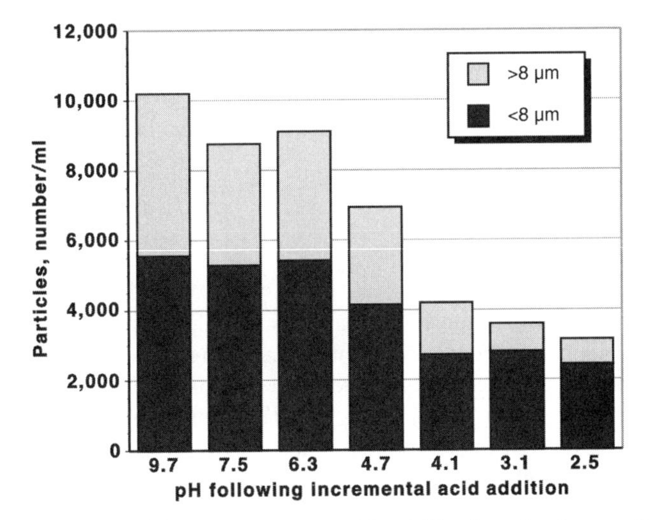

Figure 4-7 Particle count decreases with acidification of softener solids.

Dissolution of Calcium Carbonate

The series of micrographs in Fig. 4-8 illustrates the progressive dissolution of calcium carbonate (white granules on black membrane) during successive acidification. The remaining particles on the membrane, representing the residual turbidity, are translucent.

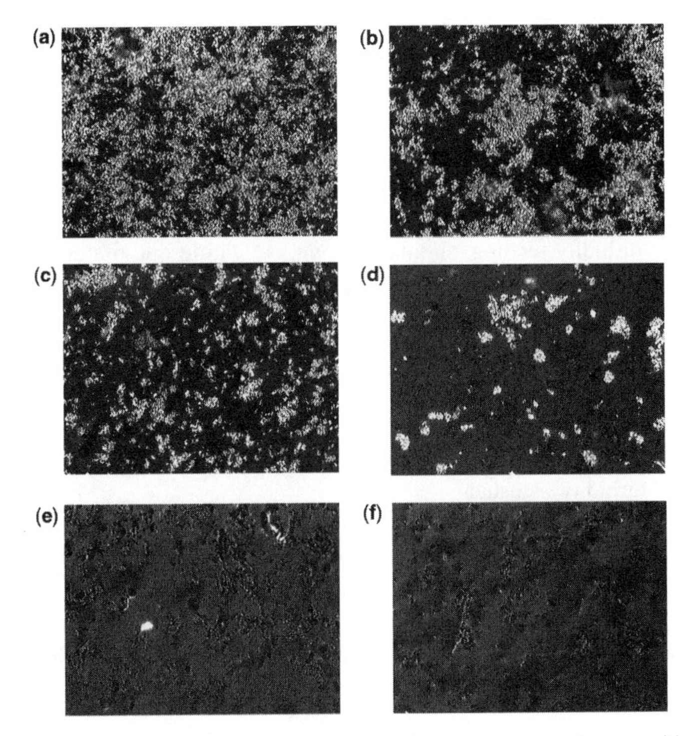

Figure 4-8 Series of micrographs illustrating dissolution of calcium carbonate with acidification.

Micrographs were taken using differential interference contrast (DIC) and cross-polarizing filters that allowed light to pass through the birefringent calcium carbonate crystals. At each pH, an aliquot was filtered to recover the solids on a 0.45 μm membrane. The membrane was subsequently *cleared* using immersion oil to create a black background for the birefringent crystals.

Effect on Native Microorganisms

A special study was undertaken to observe the effect of acid addition on organisms in the source water. Initially, organisms from water samples from Lake Bloomington were cultured in the laboratory. Four subsets of the resulting large population of live native organisms (primarily, bacteria) were subjected to varying amounts of acidification (down to pH 6.7 [the initial pH], 5.7, 4.1, and 2.0). These four subsets were placed in sample wells in a microscope slide and observed microscopically. Microvideo was taken to observe the effects of acidification on the organism's mobility. No differences in organism populations or mobility were observed at the lower pH values. This indicated that the adjustment of pH for the dissolution of calcium carbonate did not result in cell lyses that would significantly lower the turbidity caused by the native organisms present.

ACIDIFICATION PROTOCOL

Experimentation has shown that acidification of Bloomington softened water samples to pH 2.0 removes most of the turbidity caused by calcium carbonate but has no observable effect on biotic particles. Therefore, finished water sample acidification increases the meaningfulness and accuracy of turbidity as a microbiological surrogate and as a measure of actual plant particle removal performance.

For a measurement of turbidity in Bloomington finished water that is consistent with the turbidity measurement in source water, that sample pH should be adjusted to 2.0. This may be accomplished by the addition of two drops (0.1 ml) of $1+1$ (6 N) hydrochloric acid solution to a 40 ml water sample. Dissolution of lime precipitates, as indicated by a stable acidified turbidity reading, should be complete within one minute.

Therefore, the following acidification protocol has been added to Bloomington's standard method for turbidity analysis.

1. Add two drops of $1+1$ HCl to a 40 ml turbidity sample.
2. Invert several times to mix thoroughly and dissolve calcium carbonate.
3. Clean and dry the turbidity sample cell.
4. Measure turbidity as usual.

As always, the analytical protocol should observe the basic rules of turbidity measurement:

- Make sure the turbidimeter is properly calibrated with appropriate standards.
- Eliminate factors that would affect the measurement of turbidity: condensation, flocculation, sedimentation, air bubbles.

Report turbidity readings as follows:

Turbidity Range	Report to Nearest
0–1	0.01
1–10	0.10

ALTERNATE EXCEEDANCE LEVELS

It has been shown that acidification before the measurement of turbidity, as applied to a lime softening plant, results in a more accurate assessment of the removal of those particles (and potentially harmful organisms) originally present in the source water.

In these evaluations, the dissolution of lime-softening precipitates by acidification to pH 2 reduced turbidity values by 60% to 87% (i.e., only 13% to 40% of finished water turbidity might be attributable to particles other than calcium carbonate). Based on these initial observations and in the absence of acidification of the finished water samples, the designation of alternate exceedance levels of two to four times the standard exceedance levels would appear to be appropriate.

REFERENCES

O'Connor, J. T. (1990). An Assessment of the Use of Direct Microscopic Counts in Evaluating Drinking Water Treatment Processes. ASTM Special Technical Publication 1102: Monitoring Water in the 1990's: Meeting New Challenges.

O'Connor, J. T. and O'Connor, T. L. (2002). Control of Microorganisms in Drinking Water, Chapter 8: Rapid Sand Filtration. *American Society of Civil Engineers*, ISBN 0-7844-00635-9.

5

FILTER OPERATIONS

PARTICLE REMOVAL DURING FILTRATION

As part of a comprehensive water treatment system, filtration is so strongly related to previous steps in the treatment process that filtration performance cannot be looked at completely independently. Ineffective pretreatment (coagulation and sedimentation), or post-precipitation of calcium carbonate will likely result in ineffective filtration.

Filtration is often considered the principal particle removal process in water treatment. However, only a small fraction of the particles either entering a plant or produced during water treatment may be actually removed in the filters. By far, most of the particles originating in Bloomington's lake waters are removed in their softener/clarifiers. The filters are commonly referred to only as effluent *polishers*. However, federal and state regulations have caused utility operators to focus most of their attention on the performance of individual filters and filter effluents.

THE BLOOMINGTON FILTERS

Filters in the Annex Building

In total, there are 18 dual media (GAC/sand) adsorber/filters installed at the Bloomington plant. The 12 old filters (Fig. 5-1) in the 1929 Annex Building were constructed in three groups of four that went into service in 1929, 1956, and 1966, respectively. Each filter box contains 0.48 m of GAC overlying a 0.3 m layer of silica sand. Used primarily for taste-and-odor control, the GAC is replaced with new, virgin carbon on a three-year

Water Treatment Plant Performance Evaluations and Operations. By John T. O'Connor, Tom O'Connor, and Rick Twait

Figure 5-1 Annex (1929) GAC/sand filter.

schedule (Fig. 5-2). Accordingly, four filters are serviced each year. Backwash of these older filters is assisted by a surface wash system using PVC pipe and fixed nozzles projecting into the expanded bed (Fig. 5-3).

Each of the 12 filters in the Annex Building consists of a single $6.1 \times 6.6\,\text{m}$ cell, providing a surface area of $40\,\text{m}^2$ per filter. Rated at a *downward flow velocity* of 3.8 m per hour ($1.5\,\text{gpm/ft}^2$), the older filters are nominally operated around 1.5 m/h. This results in a substantial *empty bed contact time* (EBCT) of 19 minutes within the 0.48 m layer of GAC. Ignoring the volume occupied by the medium, EBCT is used as a gross measure of the process-loading rate on carbon *adsorbers*. For comparison, design EBCT is commonly 10 to 15 minutes (O'Connor, 1980).

Figure 5-2 Granular activated carbon filter cap.

Figure 5-3 Filters in 1929 Annex have fixed surface wash nozzles.

At the time of carbon replacement, the height of the sand interface is measured in each bed. Since granular filters tend to lose 5% to 7% of their media each year, make-up sand is added to restore the original sand bed height.

Filters in the Main Building

The six *new* Filters (13–18) in the Main Building were constructed in 1994 (Fig. 5-4). Each of these filters consists of dual 3.3×6.6 m cells that provide a combined area of

Figure 5-4 Main Building GAC/sand filter.

Figure 5-5 GAC cap and surface wash.

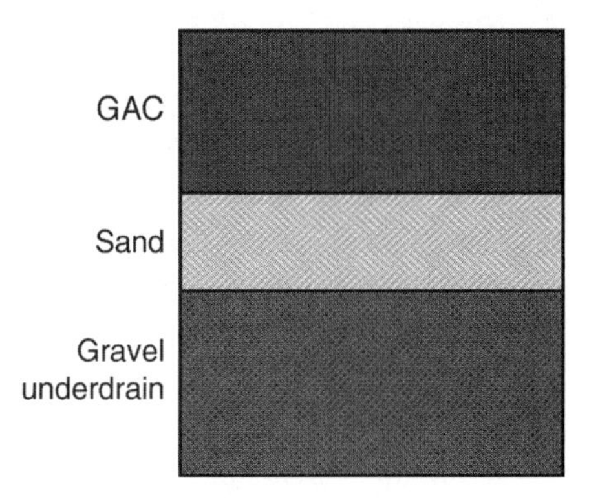

Figure 5-6 Filter media cross-section.

Figure 5-7 Removal of old GAC.

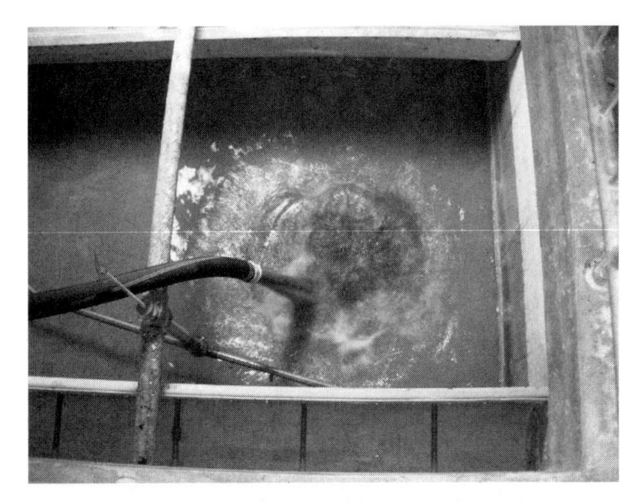

Figure 5-8 Placement of make-up filter sand.

43 m^2 per filter. They are served by a single influent and wash water gullet. A 0.6 m layer of GAC (Fig. 5-5) overlies a 0.3 m layer of filter sand (Fig. 5-6).

Rated at 7.3 m/h (3 gpm/ft^2), these filters are operated at around 4.9 m/h (2 gpm/ft^2). Through the 0.6 m of GAC, the EBCT is 7.5 minutes. Since these filters operate at a higher rate than the Annex filters, the GAC is replaced (Fig. 5-7) and sand added (Fig. 5-8) every two years.

FILTER OPERATIONS

Filter runs are limited to 48 hours before being backwashed to avoid irreversible compaction of deposits on the filter media. Backwash sequences are initiated manually in order to give operators a chance to directly observe individual filters. While observing a filter backwash, Bloomington operators look for:

- air bubbles—a sign of air binding or entrainment of air in the backwash water line;
- calcium carbonate plates and chips—derived from spalling or pressure washing of filter walls and piping;
- foreign matter in the filter—such as mudballs, cemented media, algal filaments or mats, fibers, surface accumulations;
- uneven distribution of washwater—boils, horizontal flows, lifting of media, wall effects, shrinkage cracks, separation of GAC and sand, hydraulic surges;
- media blowoff—carryover of light media (GAC, fine sand) into the backwash launders;
- unusual quantities of solids (or unusual color) in backwash water;
- foam—an indication of the presence of organic (surface-tension-lowering) compounds (Fig. 5-9).

The release of air bubbles during backwash, observed by operators at Bloomington at one time, has been controlled by installing air-release valves along the backwash water influent line.

Figure 5-9 Air bubble release causes foaming.

Filter Monitoring

Even where SCADA systems are available, operators need to closely monitor treatment plant process performance, particularly in ways that computers cannot. Automated equipment, alarms, and computers have minimal powers of observation, lack judgment, and are unable to respond to emergencies. They are also susceptible to many types of failure, such as electrical power surges and outages, plus system component failure. Corrosion and the build-up of solids in sampling lines also cause failure of monitoring systems and the recording of erroneous data due to blockages and periodic solids sloughing. Manual operation of the filters is accomplished using the filter control consoles, illustrated in Fig. 5-10.

Figure 5-10 Main Building filter control console.

Figure 5-11 Main Building filter wash water gullet.

Bloomington's filters are convenient to operate and maintain. They are completely enclosed and protected from wind-blown debris and sunlight. Individual filter banks are further enclosed within partitions having large windows for ready observation. Filter gallery areas are tiled for ease of cleaning spills and routine filter maintenance operations. The common wash water gullet, shown in Fig. 5-11, drains backwash water from both halves of the filter, which are washed in succession.

Each individual filter is equipped with a continuous flow turbidimeter. In addition to SCADA systems for recording flow rates and head loss, turbidity can be displayed visually for each individual filter.

Similarly, backwash can be automatically controlled and monitored on the SCADA control panel. *Surface wash* is set for four minutes. After surface wash is completed, *high-flow backwash* is initiated for six minutes, 40 seconds. Finally, two minutes of *low-flow backwash* complete the filter wash to ensure gentle restratification of the filter media.

IEPA Turbidity Requirements

By law, filtered water turbidities on any given filter should:

- not exceed 1 ntu*
- not exceed 0.5 ntu* after the first four hours of filter operation following a backwash

*for two consecutive 15 minutes sampling intervals.

What to Do About High Turbidities

If it appears that a particular filter is about to exceed IEPA turbidity requirements, Bloomington operators immediately take the filter out of service until the automated results can be verified and the situation properly remedied. As soon as possible, a sample of filter effluent is taken by hand for analysis using the frequently-calibrated plant operator's laboratory turbidimeter to compare with the readings being given by the continuous flow turbidimeter installed on the filter effluent line.

Continuous flow turbidimeters require frequent maintenance to avoid recording spurious turbidity readings. The small diameter finished water sampling tube leading to each turbidimeter may progressively become coated and encrusted with particles, including calcium carbonate. Subsequently, due to changes in flow, temperature, or physical disturbance of the line, these attached particles may slough from the inner surface of the tubing, causing spurious readings that erroneously indicate that the filter is passing particles.

GRANULAR ACTIVATED CARBON PERFORMANCE AND CHARACTERISTICS

Oxygen Persistence in Sand and GAC Media

It has been observed that, because of continuing microbial respiration, oxygen depletion occurs in filters that remain idle for extended periods. In addition, the pH of water in contact with the media decreases. This is consistent with microbially-mediated oxygen depletion (respiration) that results in the production of carbon dioxide. The lowered pH during periods of stagnation may result in the dissolution of calcium carbonate and magnesium hydroxide previously deposited on the filter media.

To avoid the adverse effects of stagnation, filters that have been idle for significant periods are backwashed before being returned to service.

To evaluate the effects of microbial respiration on filter media, a container of filter sand removed from a newer filter was rinsed with aerated tap water (dissolved oxygen, 11 mg/l; pH 9.0). Upon standing, the oxygen was found to become totally depleted, while the pH decreased to 8.5. The implication of these observations is that, because of continuing microbial activity, oxygen depletion will occur in filters that remain idle for extended periods.

A similar test was then conducted with both sand and GAC subsequently removed from an older filter. Whereas oxygen decreased to $7.5\ g/m^3$ in the filter sand, it decreased to $3.5\ g/m^3$ in the GAC. The pH had decreased to 8.5 and 8.0 in the sand and GAC, respectively. This would be consistent with microbial respiration that results in the production of carbon dioxide.

To conduct observations of oxygen depletion in their filters, the Bloomington laboratory has acquired a recently-developed fiber optic dissolved oxygen probe that can be inserted into the medium of a filter bed. This analytical device provides them with the capability for observing dissolved oxygen at various depths within a filter both during filter operation and stagnation. For waters that support biological growth, the degree of dissolved oxygen depletion may be a vitally-important measure of the effectiveness of filter backwashing in removing microbial accumulations.

TOC Removal by GAC and through Plant

TOC removals at the Bloomington plant have been far higher than the 25% required under the EPA surface water treatment regulations. In May 2002, most of the TOC reduction in the plant, 56%, also occurred during pretreatment (Fig. 5-12). Subsequent filtration through freshly-installed GAC further increased overall TOC removal to 77%.

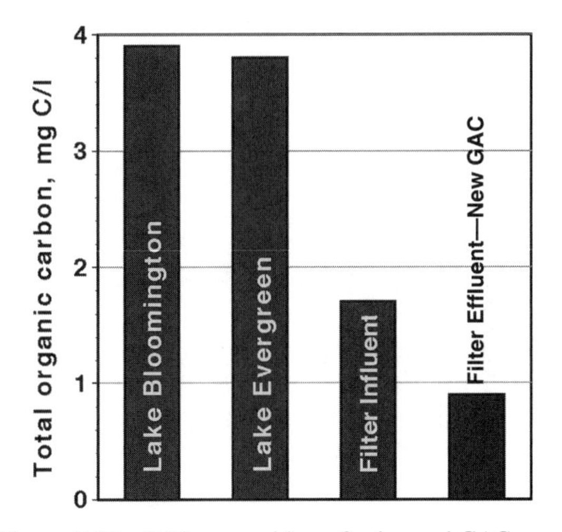

Figure 5-12 TOC removal by softening and GAC caps.

In June 2002, a comprehensive TOC profile through the Bloomington plant again showed that most of the TOC removal took place in the softeners. Still, total TOC reduction was 63%.

With the analytical equipment available in Bloomington's laboratory, it is possible to quantify the overall reduction in organic matter achieved by the GAC filter/adsorbers. This information is important for assessing the useful life of the GAC cap, as well as for determining the degree of solids removal from the GAC surface that would maximize biological activity. Where biological processes are contributing to the removal of organic matter, including taste-and-odor compounds, a perfectly clean (organism-free) filter/adsorbent surface may not be optimal. Alternately, an excessive accumulation of cell mass may lead to anoxic conditions and the generation of odorous compounds in the filter media.

Initial data on TOC [primarily, *dissolved organic carbon* (DOC)] removal in a newer filter showed that the GAC installed in the spring of 2002 initially adsorbed 50% to 60% of the TOC applied to the filter bed. Over a period of a month, as Fig. 5-13 indicates,

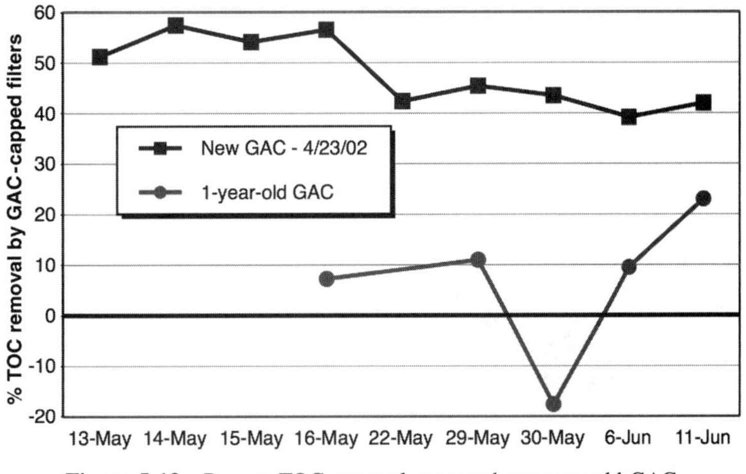

Figure 5-13 Percent TOC removal, new and one-year-old GAC.

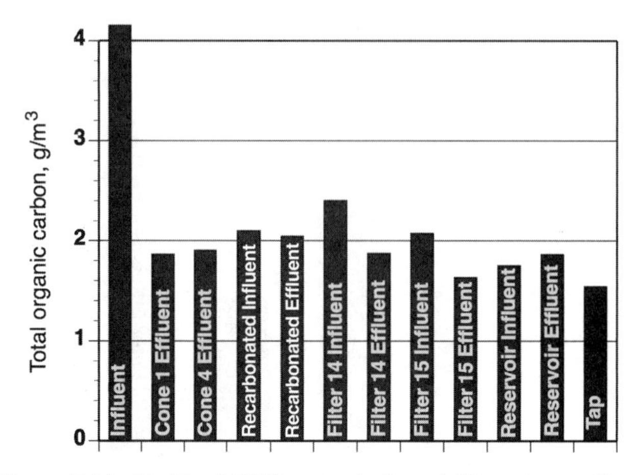

Figure 5-14 Profile of TOC removals through Bloomington Plant.

this high percentage of TOC removal diminished as the adsorptive capacity of the virgin carbon was exhausted. Alternately, a filter that contained carbon that had been in service for one year exhibited far less TOC removal. Long-term data on filters with older GAC is required to determine the degree of TOC removal achieved by biological processes alone. Figure 5-14 provides a profile of TOC through the plant on a specific date.

Changes in GAC Media

Over time, there is a noticeable discoloration (graying) of the GAC that has been in service (Fig. 5-15). Normally intensely black, the granules had begun to lighten in shade. Some of the particles among the GAC granules appeared to be solid white. These were thought to be calcium carbonate particles or chips recruited from the filter walls and piping. To determine the extent of this recruitment, a portion of GAC was dried, weighed, and acidified to dissolve the calcium carbonate. This resulted in a 2% loss of weight.

In the future, operators decided that greater care should be taken to minimize the recruitment of these calcium carbonate particles. Many appear to begin as larger platelets dislodged from the filter surfaces during pressure washing. With repeated backwash, these are broken into progressively smaller pieces. Raking of the filter surface after pressure washing should remove most of these particles.

Figure 5-15 Appearance of virgin and used GAC.

The carbon granules themselves showed varying degrees of white patches on their surface. When dislodged from the carbon by shaking in a test tube, the suspended material in the supernatant from these granules was found to consist of large numbers of microorganisms of diverse sizes and shapes. This is consistent with the characterization of GAC as a biological filter since the organic nutrient adsorbed by the carbon is converted to microbial cell mass. This cell mass is then part of the material removed by backwash during each filter cycle.

GAC Replacement

Every year, under contract with the GAC supplier, some of the GAC caps on the Bloomington water treatment plant filters are replaced with *virgin* (not previously used) carbon. The two-year-old carbon at the new plant and three-year-old carbon at the old plant are removed from each of the beds hydraulically. The carbon/finished water slurry is ejected from the bed through flexible hoses into an empty tractor-trailer capable of holding the GAC contents of a single filter bed (Fig. 5-16). The trailer returns the used carbon to the manufacturer for thermal reactivation and reuse for less critical applications, such as decolorizing sugar or rum or for waste treatment processes.

The replacement sand arrives at the plant in large totes (Fig. 5-17) that are moved using forklifts. The fresh GAC arrives in 500 kg totes that hold 3.6 m^3 of the adsorbent. The bulk (dry) density of the GAC is only 25% that of filter sand. Despite its larger particle size, this lighter material stratifies in filter beds during backwash to form a discrete layer of coarser GAC atop the denser filter sand.

The GAC and sand media in Bloomington's filters are not significantly intermixed. This discrete layering creates a sharp interface at which foreign materials and mudballs can settle and accumulate. Over time, such accumulations can partially *blind* the filter in a horizontal plane, resulting in reduced filtration effectiveness, increased rate of head loss, and decreased lengths of filter runs.

Even when freshly installed, virgin GAC will consume oxygen for some time since GAC is a strong reducing agent. GAC will also continuously consume any applied chlorine and chloramine residuals, reducing those compounds to chloride ion while the carbon itself is oxidized to carbon dioxide. This is why it is generally considered inappropriate to prechlorinate immediately before filtration through a GAC-capped filter. This consumption of disinfectant residuals largely forms the basis for the claim of *improved*

Figure 5-16 Recovery of used GAC in trailer.

Figure 5-17 Filter sand in totes.

water taste when carbon adsorbers are used for household, point-of-use treatment, as in refrigerator *filters*.

Because of its comparatively large particle size ($D_{10\%} \approx 1$ mm), GAC is not as effective a filter medium as filter sand ($D_{10\%} \approx 0.5$ mm). Whereas solids will accumulate within and on the surface of GAC, the finer particles in the filter influent are retained in the sand layer. Accordingly, the sand layer is considered critical for meeting stringent new regulatory standards on filtered water turbidity.

FILTER MEDIA SIZE DISTRIBUTION

The following comparative evaluations were conducted on the filter media while GAC was being removed and replaced in filters 13, 15, and 17 of the new plant and filters 2, 3, and 4 of the old plant in the spring of 2002.

GAC Size Distribution and Characteristics

From visual inspection, the GAC and sand media removed from filters that had been in service for two or three years appeared very different from the fresh media that was being installed to replace the GAC and augment the sand. To confirm the visual observations, media size distributions were determined using laboratory sieve equipment (Figs. 5-18 and 5-19).

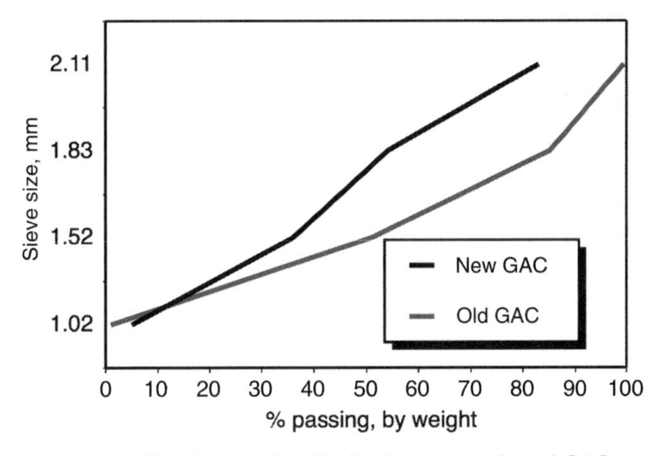

Figure 5-18 Particle size distribution, new and used GAC.

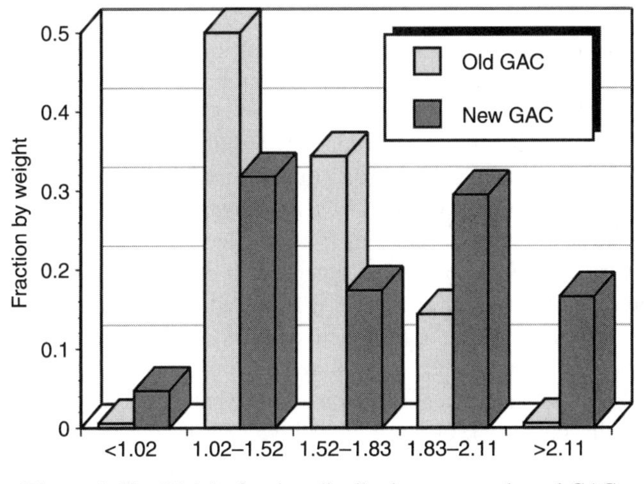

Figure 5-19 Weight fraction distribution, new and used GAC.

From the sieving results obtained, it was clear that the size distribution of the GAC that had been in use for two years had narrowed considerably. This is partly attributable to the abrasion of the larger GAC granules so that their size is reduced. At the lower end of the size range, the finer GAC (<1 mm) granules may have been carried out of the filter during backwash. The result is a more uniform GAC size distribution. Despite the decrease in *uniformity coefficient* (UC), the *effective size* (ES or $D_{10\%}$ particle size) appeared to have remained slightly above 1 mm. Overall, a narrowing size distribution with age makes the GAC, progressively, a more uniform and, from a hydraulic standpoint, ever more desirable filter medium.

Sand Size Distribution and Characteristics

From the sampling of the underlying sand layer conducted during the changeover of the GAC, it is evident that the size and weight distribution of the filter sand also has

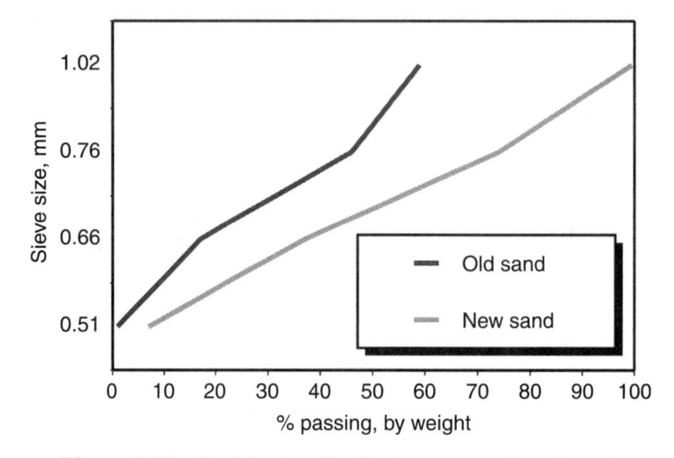

Figure 5-20 Particle size distribution, new and used sand.

changed since it was first installed (Figs. 5-20 and 5-21). To begin with, the $D_{10\%}$ size appears to have increased from approximately 0.52 to 0.58 mm. In addition, the $D_{60\%}$ size increased substantially from about 0.72 to 1.05 mm.

The ratio $D_{60\%}/D_{10\%}$, encompassing 50% of the medium by weight, is used to calculate the UC. The smaller the UC, that is, the more uniform the grading, the more desirable the material is as a filtration medium. Partly due to intermixing of media, the UC in the filter sand layer has increased from 1.38 to 1.81. This increase in size distribution with use is evident from the relative appearances of the old and new sand (Fig. 5-22).

These size increases will also combine to make it increasingly difficult to attain the degree of expansion of the sand layer that is believed to be optimal for particle scour and cleaning of the medium. For washing sand filters, without the assistance of surface wash, air, or auxiliary scour, a 20% to 50% bed expansion is commonly used. Since the surface wash at Bloomington does not reach into the expanded sand layer, sand bed expansion should be in this range. The absence of well-defined stratification within the sand layer may be an indication that this degree of expansion is not being achieved.

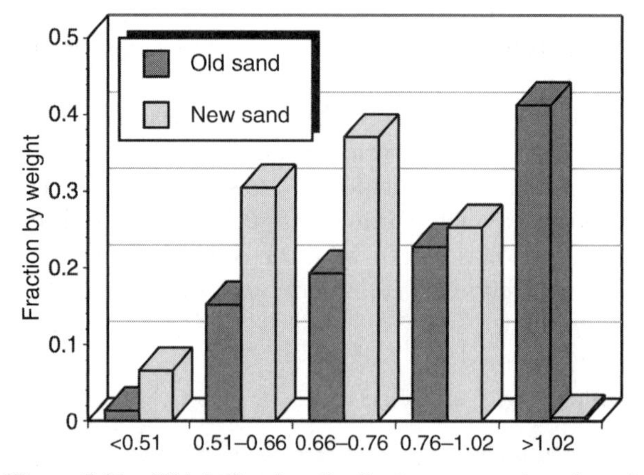

Figure 5-21 Weight fraction distribution, new and used sand.

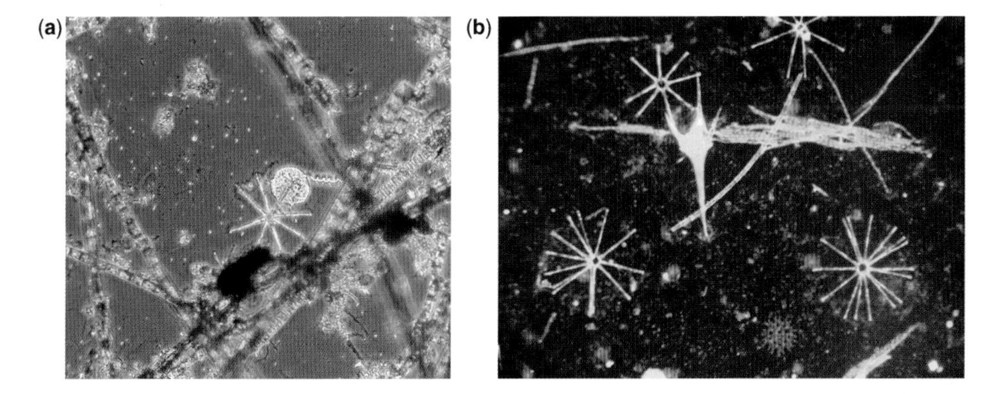

Figure 1-3a–b (a) Filamentous algae, diatoms, bacteria, and organic debris in Lake Bloomington water. Micrograph illustrates the wide distribution of particle sizes and shapes in lake water. Large, fluorescing particles are algae and diatoms (10 to 40 μm); smaller fluorescing dots are, primarily, bacterial cells (1 μm); (b) Micrograph of Lake Bloomington water reveals fragile diatoms (Asterionella), algal filaments, numerous bacterial cells, and loose particle aggregations. The flocculent extracellular excretions of microorganisms may entrain numerous smaller particles. Loose and fragile aggregations may be disrupted by passage through capillaries of particle counting devices.

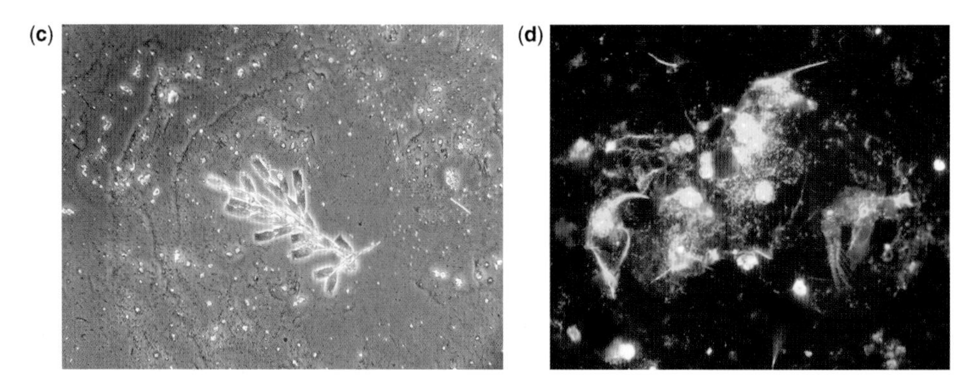

Figure 1-3c–d (c) Dinobryon colony in lake water; (d) Disintegrating *Daphnia* in lake water colonized by numerous bacterial cells (small orange dots).

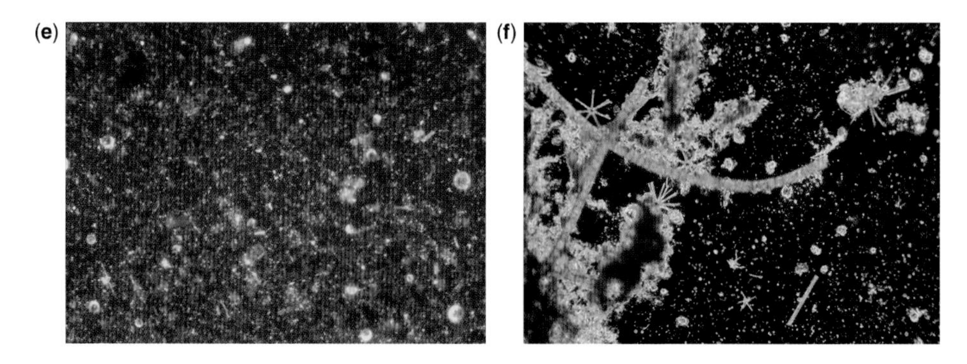

Figure 1-3e–f (e) Turbid Evergreen Lake water with inorganic solids and abundance of bacterial cells; (f) Evergreen Lake water with aggregations of debris attached to algal filaments. Smaller particles appear to be flocculated by extracellular polymers secreted by algae.

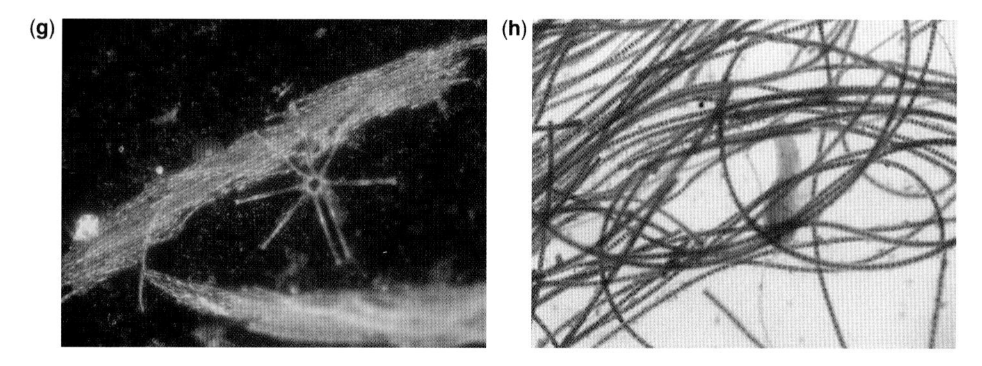

Figure 1-3g–h (g) Sheaths of filaments contain large numbers of algal cells; (h) Mat of algal filaments entrain grazing zooplankter.

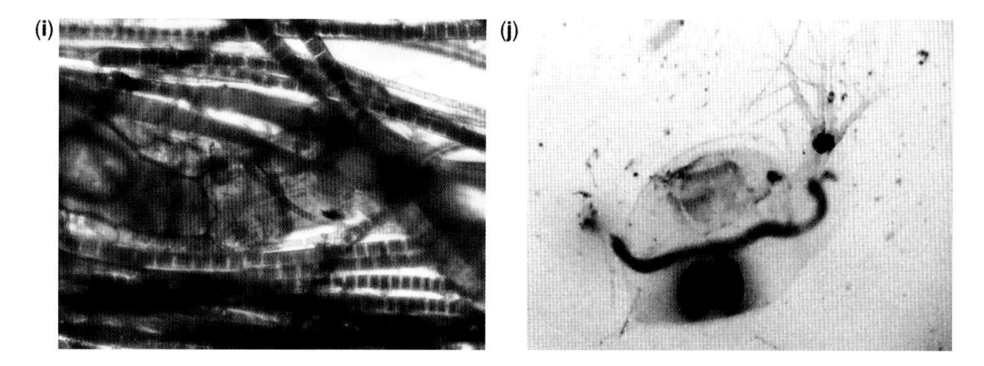

Figure 1-3i–j (i) Algal filaments showing well-defined cells enclose macroinvertebrate; (j) Fecal matter in gut of *Daphnia*. Small, motile aquatic crustaceans with entrained microscopic organisms passing through treatment constitute *particles of potential health significance*.

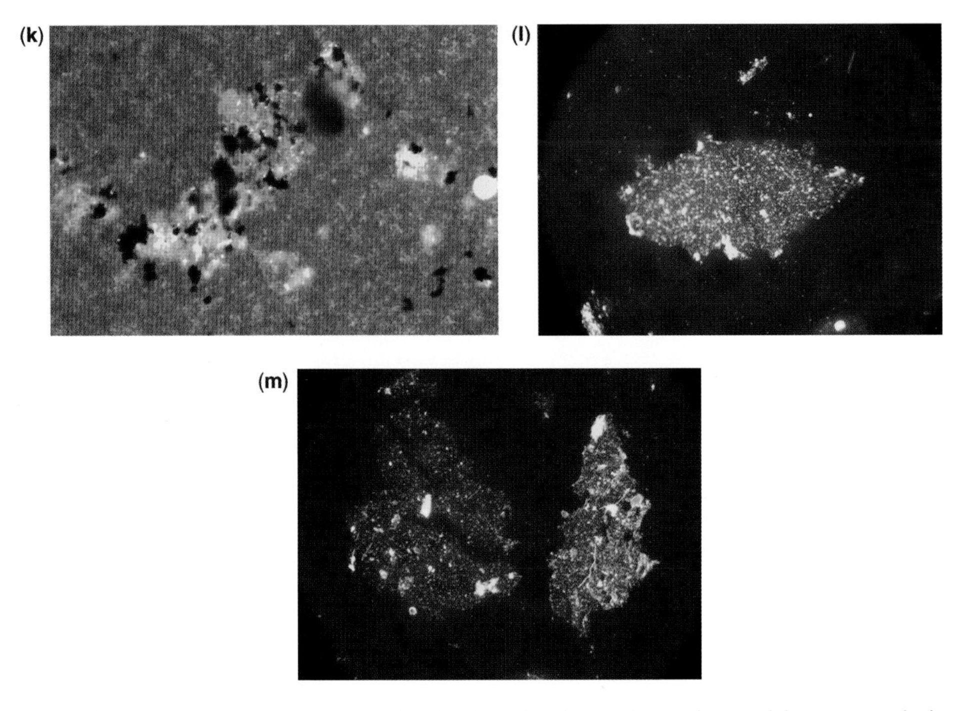

Figure 1-3k–m (k) Application of powdered activated carbon and coagulants to lake water results in formation of flocs with embedded carbon (black) and microorganisms (yellow); (l) Well-defined floc formed in softener shows near complete entrainment of micrometer-sized particles; (m) Fragile flocs in softener effluent show entrained bacteria, algal filaments, inorganic debris, and precipitates formed during softening in a matrix of magnesium hydroxide.

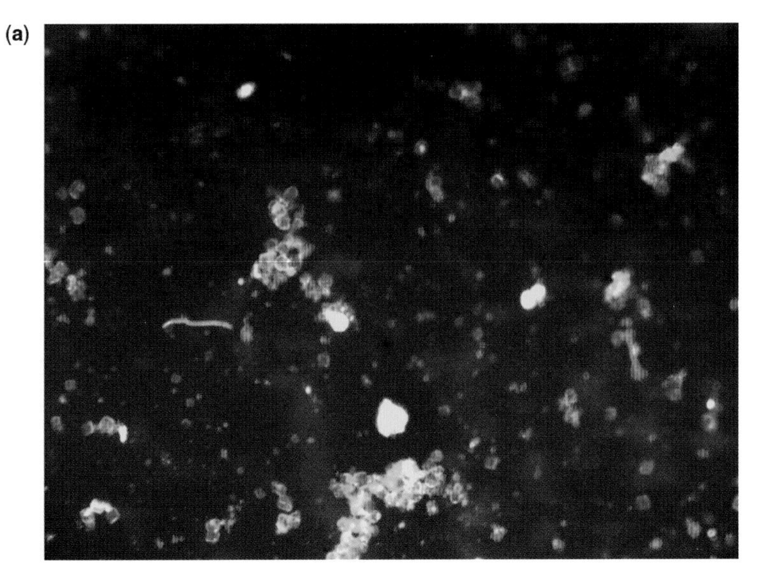

Figure 1-7a Calcium carbonate crystals (green) dominate particles found in the recarbonation basin and, hence, filter influent.

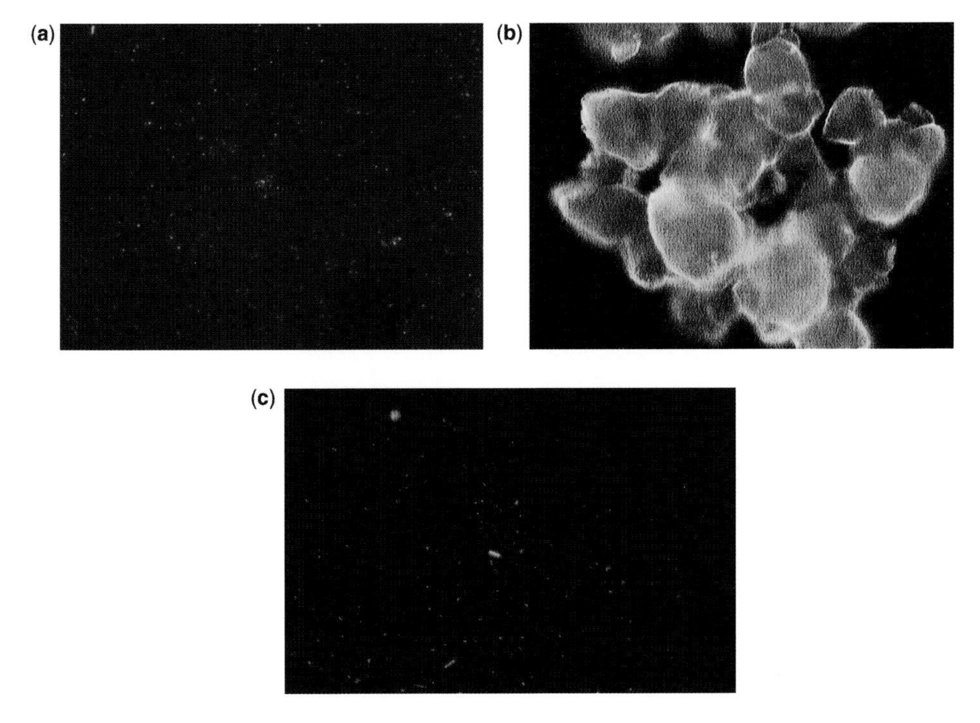

Figure 1-8a–c (a) Planktonic bacterial cells predominate in filter effluent. Unprotected bacterial cells are believed to be most susceptible to action of disinfectants; (b) Cluster of calcium carbonate crystals are occasionally found in filter effluent. Crystals in filtered water do not appear to harbor microorganisms; (c) Discolored (stressed) bacterial cells observed in filtered water following final disinfection.

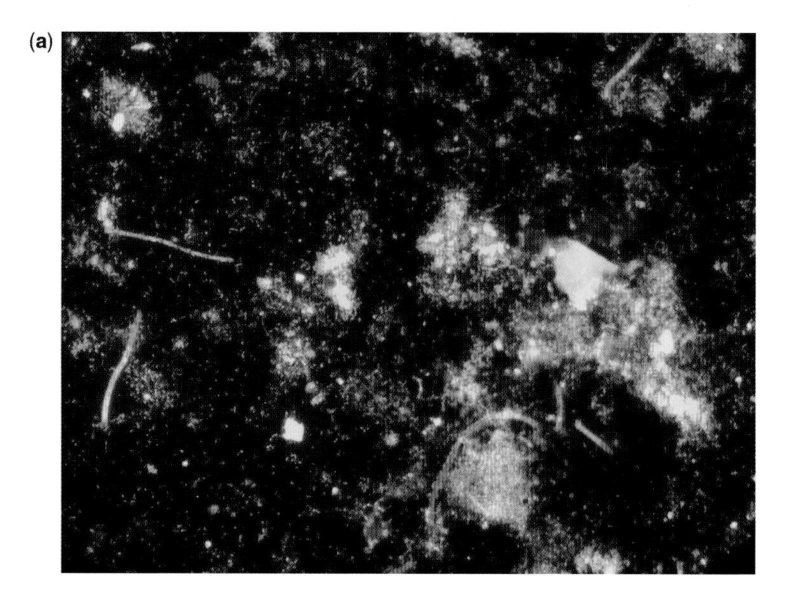

Figure 4-3a Material rinsed from filter media indicate abundance of attached growth of bacterial cells and filaments largely embedded in slime.

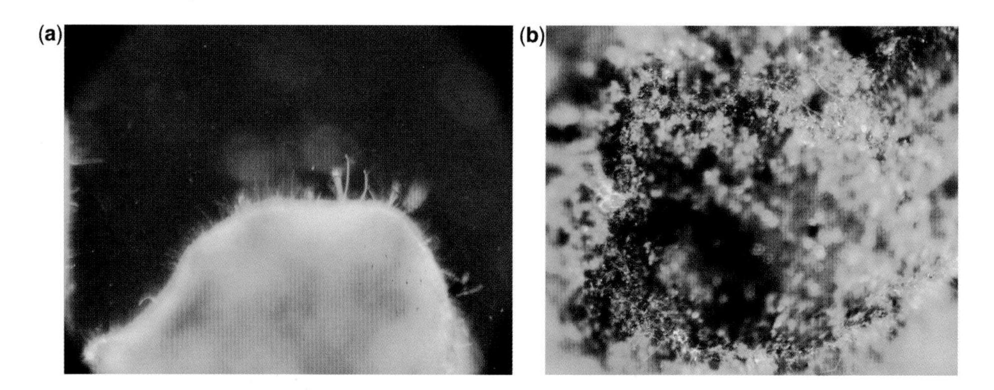

Figure 5-23a,b (a) Bacterial filaments extending from GAC granule surface. Stalked bacteria that project from surface may aid in the attachment of microscopic particles to GAC granules. (b) Bacterial colonies on GAC granule surface. Patchy microbial growth on the surface of GAC granule utilize adsorbed organic substrates.

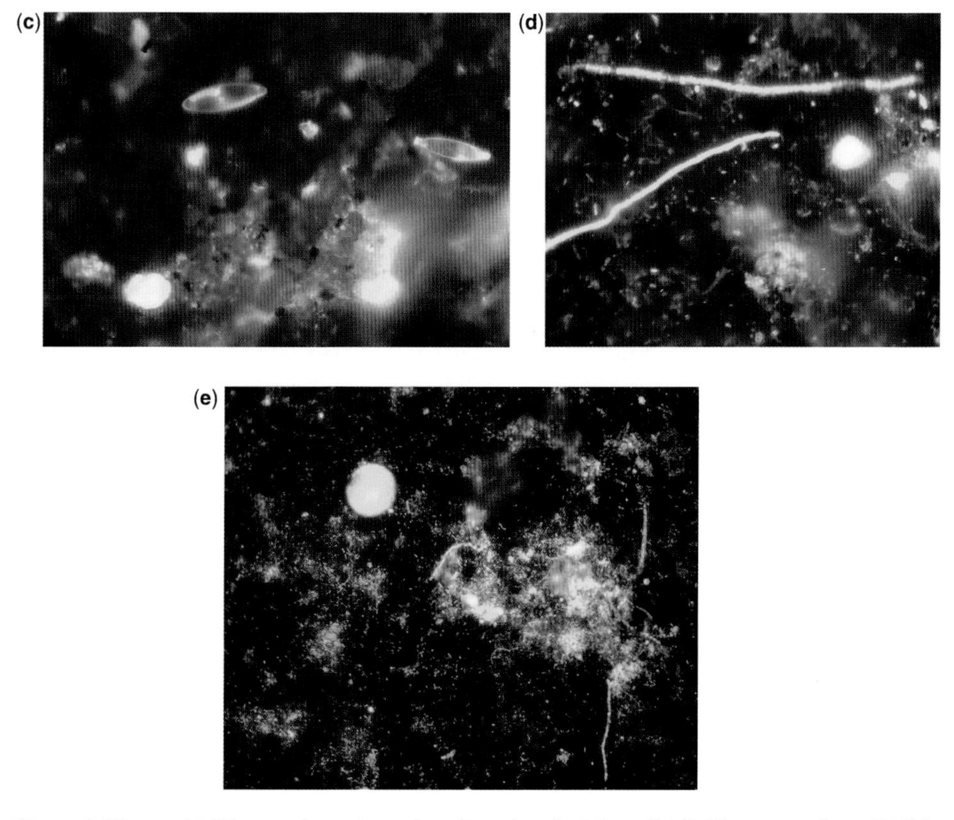

Figure 5-23c–e (c) Diatoms, bacteria, carbon fines detached from GAC. Rinse water from GAC in service for an extended period reveals a large diversity of organisms as well as carbon fines. (d) Bacterial rods and filaments detached from GAC. Rinse from GAC indicates growth (cell division) of large bacterial cells and formation of long filaments. (e) Bacterial cells and filaments dislodged from filter sand. Although less than from GAC, rinse water from sand shows substantial accumulations of microorganisms and aggregations in detached slime.

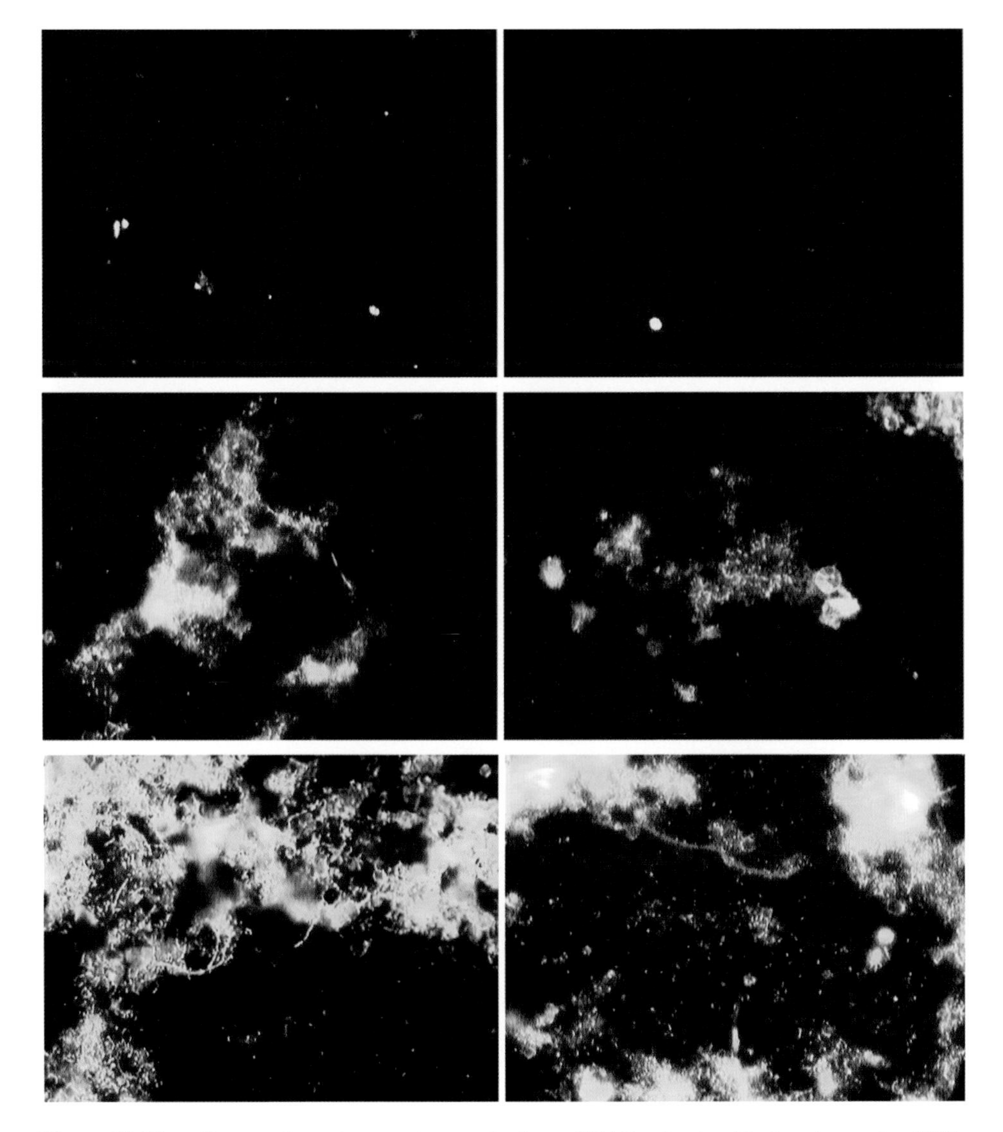

Figure 12-10a Comparative microscopic examination of GAC extracts with time in service (400x magnification). Row 1. Few fluorescing particles were observed on the extract from virgin GAC. Row 2. After one year, numerous bacterial cells, filaments, algae, diatoms, $CaCO_3$ crystals in gelatinous matrix. Row 3. After two years, abundance of unicellular and filamentous bacteria in gelatinous matrix; possible senescent filament.

Figure 5-22 Appearance of new and old filter sand.

Visual observation of the sand removed from the filters at various depths (Fig. 5-23) indicates that the increase in effective size (ES) and uniformity coefficient may have occurred for the following reasons:

- Due to hydraulic surges, *torpedo* (larger, heavier) sand, which forms the stable base for the filter sand, was thrust upward into the filter sand layer.
- Some larger particles (white chips) of calcium carbonate, dislodged during filter box maintenance, had migrated through the GAC layer and settled onto the top of the sand layer.
- A portion of the finest filter sand (<0.5 mm) was absent, indicating that it may have migrated and escaped down into the underdrain system.
- Some GAC was found in the sand sample. While this was picked out as much as possible before the sand sieve analysis was conducted, these larger GAC particles would tend to increase the apparent effective size of the sand mixture.

Figure 5-23 Appearance of sand removed from top and bottom of layer.

Despite the observed changes, the sand medium remains in the range commonly utilized for filter sand, though at the upper boundary. The addition of make-up sand, practiced as part of routine GAC replacement, will lower the existing size distribution. Future additions of make-up sand may utilize still finer-grained material (e.g., ES 0.4 mm) to restore the sand to its initial size range.

Alternately, the addition of three to four inches of a still finer (ES = 0.2 to 0.3 mm) and denser garnet sand (specific gravity = 4.0 as opposed to 2.6 for silica sand) may serve to intermix with the existing media and further increase filtration efficiency.

Ultimately, however, the regulated test of filter media particle removal effectiveness is the measurement of turbidity (light scattering). Continuous monitoring of filtered water turbidity profiles, as currently practiced by Bloomington Water, demonstrates that, under normal circumstances, filtration efficiency routinely exceeds regulatory requirements by a significant margin.

MICROBIAL GROWTH ON FILTER MEDIA

In the absence of the application of disinfectant before filtration, the significant surface area provided by the fine-grained filtration media offers ample opportunity for microbial colonization and growth. Although this phenomenon has been observed in water treatment practice from earliest times, it has been rediscovered as *biological treatment* or *biological filtration*.

A number of benefits can accrue to the development of a microbial community on the filter media. A portion of the organic content of the filter influent, the *labile* or *biologically assimilable* portion, is converted to cell mass and removed. While this may only be a small fraction of the dissolved organic carbon remaining after softening, these labile compounds may include those contributing to tastes and odors.

Accordingly, the maintenance of an active, aerobic microbial community on the filter media should prove beneficial, particularly during episodes when Bloomington's lake waters become enriched with nutrients, leading to algal growth and the production of taste-and-odor compounds.

Subsequently, *aerobic respiration* by accumulations of biological growth can lead to the loss of oxygen and anoxia within the filter bed. Periodic sloughing of attached accumulations of organism cell mass can lead both to turbidity excursions and elevated numbers of organisms contributing to heterotrophic plate counts (HPC).

The micrographs of Figs. 5-23a and 5-23b illustrate the abundance of green-fluorescing microbial filaments and stalked bacteria colonizing GAC granules from a filter removed from service at Bloomington. The protruding filaments (attached, stalked growth) may assist in the attachment of particles from the filter influent to the filter media.

Particles Washed from GAC

A portion of the accumulation of organisms on the colonized GAC is stripped off the granules with each succeeding filter backwash. Their removal reflects the removal of the assimilable portion of the dissolved organic carbon achieved by the filter.

The micrographs (Figs. 5-23c and 5-23d; see color insert) illustrate the particles and organisms removed during backwash. These micrographs show a large and diverse group of rod-shaped bacteria (green dots), generally embedded in clumps of detached slime

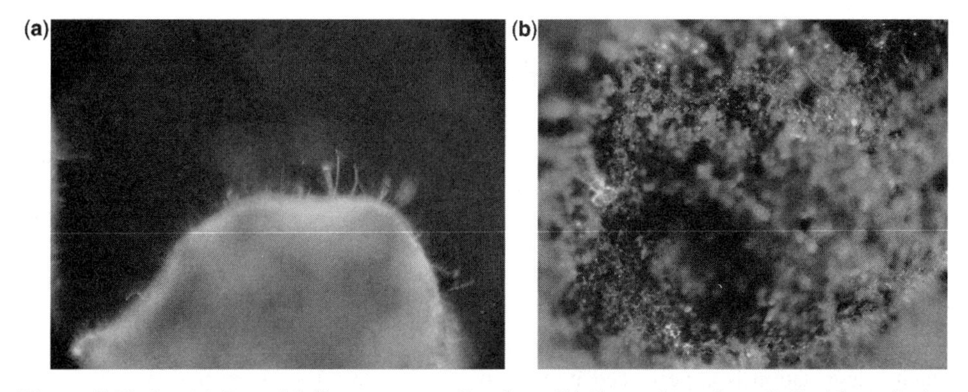

Figure 5-23a,b (a) Bacterial filaments extending from GAC granule surface. Stalked bacteria that project from surface may aid in the attachment of microscopic particles to GAC granules. (b) Bacterial colonies on GAC granule surface. Patchy microbial growth on the surface of GAC granule utilize adsorbed organic substrates. (See color insert.)

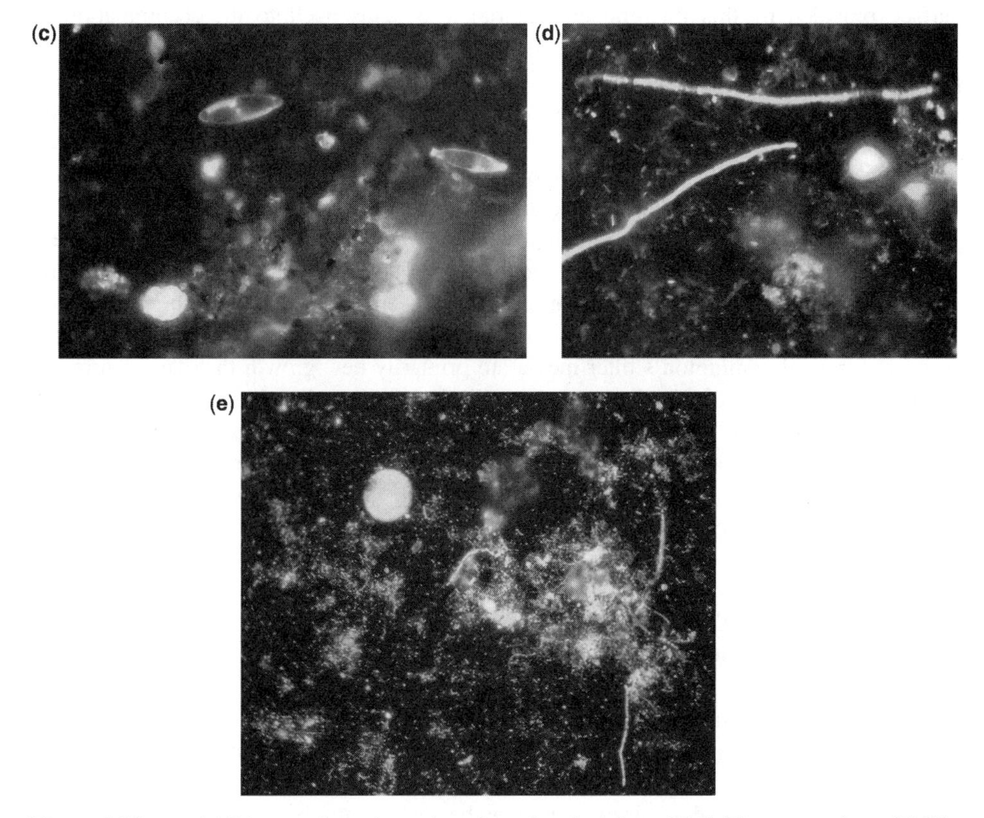

Figure 5-23c–e (c) Diatoms, bacteria, carbon fines detached from GAC. Rinse water from GAC in service for an extended period reveals a large diversity of organisms as well as carbon fines. (d) Bacterial rods and filaments detached from GAC. Rinse from GAC indicates growth (cell division) of large bacterial cells and formation of long filaments. (e) Bacterial cells and filaments dislodged from filter sand. Although less than from GAC, rinse water from sand shows substantial accumulations of microorganisms and aggregations in detached slime. (See color insert.)

(*extracellular polymer*) along with carbon fines (black). Long bacterial filaments and diatoms are also present.

The production and subsequent removal of the labile portion of the DOC by this diverse group of organisms in the GAC-capped filter may also contribute to the reduction of microbial growths in the distribution system. These organisms have also been credited with oxidizing haloacetic acids (HAA) formed by prechlorination before filtration.

Alternately, the presence of a large microbial community has spurred studies of the degradation of water quality in filter beds when they are taken out of service for 6 to 24 hours. As noted, oxygen is found to decrease within the filter pore water within hours. While DOC, ammonium, and nitrite ion concentrations in the filter pore water then increase, attached microbial cell mass declines under the anaerobic conditions. As a result, water quality degradation is experienced upon start-up. To avoid this deterioration and restore the DOC biodegradation performance of the filters, idled filters are backwashed before being returned to service.

Particles Washed from Sand

While silica sand is inert and does not provide nutrient for microbial growth, the extensive surface provided by this fine-grained material also lends itself to the accumulation of attached microbial growth. Because of this, in many treatment systems, the influent to sand filters is *prechlorinated* to prevent the accumulation of microbial slime on filter media.

Figure 5-23e shows organisms recovered in the water rinsed from Bloomington's filter sand. While not as productive as GAC, the microbial community on sand is abundant and appears diverse. Many of the bacterial cells (orange-red rods and filaments) appear aggregated or embedded in clumps of detached slime.

In viewing these micrographs, it is important to remember that any organisms of potential health concern that were present in the lake water source have largely been physically removed by coagulation and sedimentation or have been inactivated at the high pH maintained during the lime softening process. Instead, the organisms shown in the micrographs recovered from Bloomington's filter media are primarily new growth that has occurred as the result of surface colonization and the utilization of the dissolved organic matter in the filter influent. These organisms are not known to represent a human health hazard. For example, similar results may be observed wherever individual consumers use household treatment devices (e.g., undersink, faucet, cartridge, or refrigerator filters) containing granular activated carbon.

FILTER EFFLUENT: CALCIUM CARBONATE POST-PRECIPITATION

While filtration results in the removal of most of the filter influent calcium carbonate, a continuing, slow post-precipitation may result in the formation of calcium carbonate particles within the filter beds and on the filter media. As a result, finished water turbidities may increase over normal levels. This turbidity may be almost entirely due to calcium carbonate (Fig. 1-6b).

Each of Bloomington's 18 filter units is continuously monitored by dedicated flow-through turbidimeters located on each filter effluent pipe. This data is sent to the control room electronically by a SCADA system that allows the operator in charge to observe

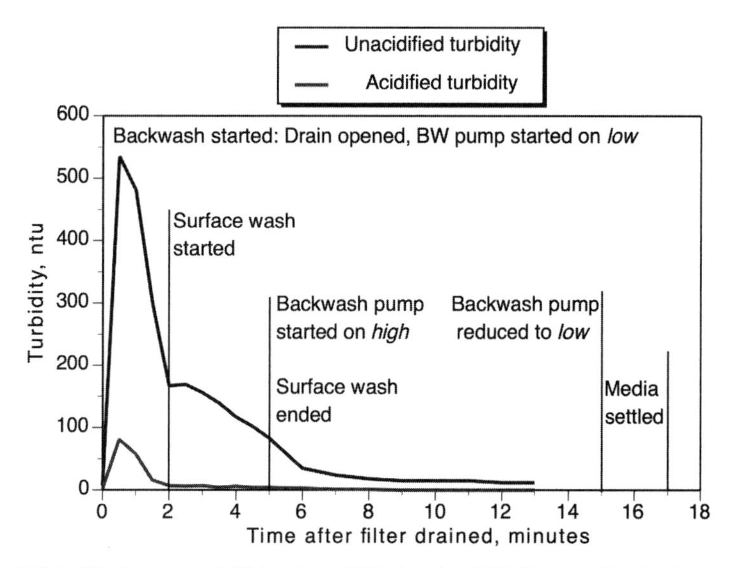

Figure 5-24 Washwater turbidities (unacidified and acidified) through a backwash cycle.

the performance of each filter on a real-time basis. If a turbidity excursion (>0.3 ntu) is observed in any one of Bloomington's filters, the operator takes an effluent sample manually from that filter. This grab sample is acidified according to Bloomington's acidification protocol and a precise turbidity measurement is made on the laboratory turbidimeter in the operational control laboratory.

The plot shown in Fig. 5-24 for acidified turbidity illustrates the predominance of calcium carbonate in the solids removed from the filters during backwash. After acidification, only an average of 7% of the unacidified turbidity remained.

FILTER WASHING

Filter Washing Procedure

At Bloomington, filters are backwashed every 48 hours unless filter effluent turbidity has exceeded 0.5 ntu or head loss is greater than 7 feet, both rare events that trigger separate alarms. Figure 5-24 depicts the water turbidities observed as a filter backwash proceeds.

At the filter control console, filter backwash begins with pressing *Start* to initiate the backwash sequence. This closes the *influent valve* to the filter being washed. With the closing of the influent valve, both the right and left sides of the filter are drained. The drainage of the filter to below the medium surface can be observed visually.

After draining, the *effluent valve* is closed and the *drain valve* opened. The low backwash pump is then started against the (closed) *low* backwash valve. The low backwash valve opens only after the pump has started. This is time 0 in Fig. 5-24.

Two minutes after low backwash has started and the medium has been hydraulically lifted over the static surface wash arms, the *surface wash* is activated and the arms start to rotate. A slight increase in backwash water turbidity is observed at this time. The surface wash will continue for approximately three minutes, after which high-rate backwash is initiated.

At five minutes, the high backwash pump is started against the (closed) *high* backwash valve and the medium is further expanded with the opening of that valve. A high-rate wash of 122 m/h (50 gpm/ft^2) is then used to expand the sand layer 20% to 50% for ten minutes. The duration of the high-rate wash may be modified by the operator for the seasonal (temperature) and pretreatment conditions. For example, following the addition of polymeric coagulant aids, the backwash is often extended to assist in removing the more tightly attached polymer.

Two minutes before completion of high-rate backwash, the pump is returned to low-rate wash. This feature allows gradual separation and settling of the sand and GAC media to reestablish stratification.

With completion of backwash on the left side, the entire wash cycle is repeated on the right side before returning the filter to service.

Effectiveness of Backwash in Removing Bacterial Accumulations

Studies were conducted as part of an effort to observe the effectiveness of backwash in removing bacterial accumulations from filter media. To start, luminescent probes were used to observe DO depletion rates in BOD bottles containing GAC rinse water suspensions. Measurements of dissolved oxygen depletion in suspensions derived from GAC extracts (before and after filter backwash) illustrate the effects of bacterial removal. As indicated by the slopes of the lines, the DO depletion rate declined from 1.02 g/m^3/h prior to backwash to 0.83 g/m^3/h following backwash.

These results, although preliminary, indicate that this innovative procedure can provide an index of the degree of removal of bacterial accumulations on filter media. If greater bacterial removals from more vigorous or extended backwash were found to result in higher (or lower) removals of taste-and-odor-producing compounds, backwash procedures would be modified.

Measurement of Filter Bed Expansion during Backwash

A pan flute device was fabricated in the Bloomington plant shop to determine bed expansions during backwash. Removable pipe sections allow this one device to be used in both the old plant where bed expansions are low and in the new plant where bed expansions are high.

For filters in the old plant, expansion of the 0.48 m GAC layer was anticipated to be 30% (or 0.14 m) at the warmest water temperature of 30°C and an upflow rate of 0.8 m/min (20 gpm/ft^2). A concurrent 10% expansion of the 0.3 m filter sand layer would add another 0.03 m to the bed expansion for a total of approximately 0.17 m.

Instead, as measured using the pan flute device, total bed expansion during a filter backwash was found to be only 0.09 m. This degree of expansion would constitute a low-rate wash resulting in only relatively weak cleansing of the filter media. This result was qualitatively confirmed by the subsequent rinsing of the GAC samples collected from the surface of a freshly-washed filter. The old filters are currently limited to this low expansion because the backwash water pumps lack capacity to provide greater flow to further suspend the media.

Since, over the years of service, some of the filter media has changed in size and some has been lost in operation, it would be instructive to conduct laboratory column studies using filter media removed from several typical filters to determine the backwash rates that are, in fact, effective in achieving the desired degree of expansion and cleaning.

These pilot columns can subsequently be used to observe the effect of polymer addition on filter performance, head loss, and removal of attached solids.

PROCEDURE FOR EVALUATING BIOLOGICAL ACTIVITY ON FILTER MEDIA

- Collect surface samples of filter medium immediately before and after backwash.
- Gently compact 50 ml of wet medium into measuring container.
- Transfer medium into BOD bottle along with 100 to 200 ml of rinse water (membrane-filtered filter effluent).
- Cap and shake BOD bottle vigorously for two minutes to strip attached solids from medium.
- Add aerated rinse water to fill BOD bottle and mix to blend in with rinse water.
- Add magnetic stirring bar to BOD bottle.
- Place bottle on magnetic stirrer set on lowest mixing rate to maintain solids in suspension.
- With cap off, insert oxygen probe into bottle and take initial measurement of DO and temperature.
- Take measurement every 5 to 15 minutes depending on rate of DO depletion.
- After 50% of oxygen is consumed, plot line of DO versus time.
- From linear slope of plot, estimate g/m^3 DO depleted/hour.

RECOMMENDED TESTING OF AUXILIARY AIR SCOUR

Auxiliary air scour at 0.6 to 1.5 m/min (2 to 5 $ft^3/min/ft^2$), along with low-rate water wash, is known to provide more effective cleaning action than water wash alone. While auxiliary air scour can reduce backwash water requirements, the increased turbulence it creates can also result in disruption of supporting layers of gravel.

Accordingly, Bloomington is considering pilot testing a replacement for a cast iron filter underdrain system in one filter at the old plant. A porous stainless steel filter bottom would replace the current pipe lateral underdrain and allow for placement of the filter sand directly atop the new underdrain. This system would accommodate air scour and reduce the overall depth of the current filter bed by approximately 0.5 m. This would also allow for an increase in the depth of the present GAC filter cap to about 0.76 m, providing greater empty bed contact time for removal of taste-and-odor-producing compounds while still allowing more head space for bed expansion during backwash. However, fine pores in the underdrain system to be tested have shown the tendency to clog and require maintenance, particularly when filtering lime-softened water.

It has been estimated (in 2006) that a test filter refurbishment could be installed by the plant personnel at a cost of $54,000 using virgin GAC that is already stored on the Bloomington plant site. Comparative data from the operation of this test filter with the other old plant filters could quantify the benefits to be realized if all the old plant filters were subsequently rehabilitated. An increase in filtration capacity might be one such benefit since pipe laterals introduce relatively high head loss.

To provide stratification of the dual media bed, a high rate water wash 0.6 to 0.9 m/min (15 to 23 gpm/ft^2) would have to follow the cessation of air scour. Reportedly, air scour can also be expected to virtually eliminate the mudball formation that has been observed at the media interface in some of the old plant filters.

The replacement underdrain system under consideration promises:

- Uniform distribution of backwash water
- Rapid installation
- Integral air scour
- Stainless steel construction
- Low underdrain profile

REFERENCES

AWWA, ASCE (1998). Water Treatment Plant Design, Third Edition, McGraw-Hill.

O'Connor, J. T. et al. (1980). Removal of Trace Organics from Drinking Water Using Activated Carbon and Polymeric Adsorbents, MERL, ORD, USEPA, Cincinnati, OH.

O'Connor, J. T. et al. (1985). Chemical and Microbiological Evaluations of Drinking Water Systems in Missouri. *Proc. AWWA Annual Conf.*, Washington, D.C.

O'Connor, J. T. and O'Connor, T. L. (2002). Control of Microorganisms in Drinking Water, Chapter 8: Rapid Sand Filtration. *American Society of Civil Engineers*, ISBN 0-7844-00635-9.

6

GRIT REMOVAL

The proprietary ClariCone, manufactured by CB&I, is an upflow slurry blanket contact softener/clarifier that encompasses the lime softening, coagulation, and settling in an inverted cone-shaped basin. Influent water and lime slurry are introduced tangentially to the cylindrical bottom section of the tank (Fig. 6-1). As this mix swirls and rises in the conical section of the tank, its upward flow velocity decreases progressively owing to the increasing cross-sectional area of the inverted cone.

In the case of Bloomington, the softening reaction results in the precipitation of both calcium carbonate and magnesium hydroxide. The magnesium hydroxide serves as a coagulant that entrains both particles in the influent water and calcium carbonate. At design flow and at approximately 9 m (30 ft.) above the cylindrical base of the tank, the upward flow velocity decreases to 3.7 m/h (0.2 ft/min), the approximate settling rate of the floc formed.

Where the floc settling velocity equals the rise or overflow rate of the liquid in the tank, a *slurry blanket interface* is formed. Above the blanket, the water is progressively clarified as its rate of rise continues to decrease in proportion to the square of the cone diameter. The clarified effluent is then collected in a single, large, adjustable, horizontal trough where the overflow velocity has declined to 2.7 m/h at design flow.

Solids are collected from the softener at two discrete levels. A portion of the light, coagulated solids from the top of the blanket enters an internal cone-shaped *slurry concentrator*. Some solids compaction occurs within this concentrator and, periodically, its contents are withdrawn through a *slurry concentrator discharge line* at the base of the tank.

In addition, there is an accumulation of heavier solids, including *grit* and unreacted lime, with settling velocities >1.5 m/min, in the cylinder at the base of the tank. This cylindrical *grit settling compartment* is the part of the cylinder (*can*) below the bottom of the influent pipe, where dense material will collect and compact.

Water Treatment Plant Performance Evaluations and Operations. By John T. O'Connor, Tom O'Connor, and Rick Twait

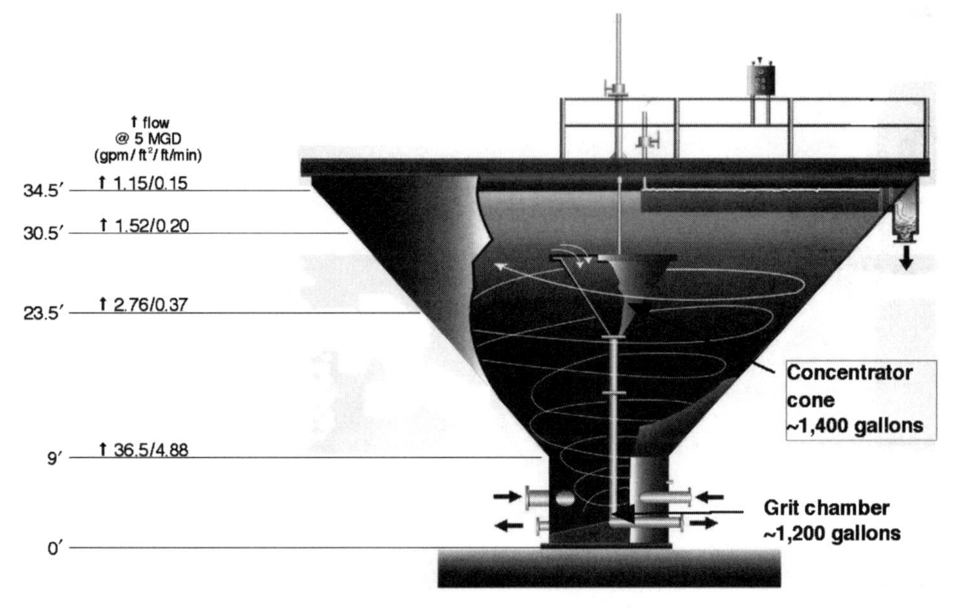

Figure 6-1 Illustration of ClariCone operation (courtesy of CB&I).

GRITTING

Purpose of Gritting

Gritting is the process of removing the settled material from the lower cylinder of the softener. If an excessive amount of heavy material (unreacted lime, calcium carbonate, sand, silt, chert, heavy debris from influent water) is allowed to accumulate in the bottom of the cylinder, the resulting mound of material can disrupt the helical flow of influent water and lime slurry, deflecting the flow upward and causing short-circuiting.

Material and Source

Much of the grit is acid-insoluble material that is present in the product lime itself. It consists of impurities, such as silica sand and chert, as well as dead-burned (unreactive) lime. Since approximately 145 g/m^3 of lime containing as much as 5% to 8% grit are applied to a softener treating 19,000 m^3/d (5 mgd), about 138 to 220 kg/d of grit may be introduced along with the lime feed.

Materials accumulated in the settling compartment are also composed of heavier particles from the influent water, as well as the bulk of material precipitated from the influent water due to lime addition. Since each kilogram of lime added results in approximately 2 kg of precipitate, an additional 5,500 kg/d of solids are generated from the water treated.

At a rated flow of 19,000 m^3/d, the average upward water velocity in the bottom cylinder of the softener is 1.5 m/min. Therefore, in order to settle into the cylinder, a particle must have a greater settling velocity.

Grit Storage

The volume of the space where grit might accumulate was estimated from two slightly different design drawings. Results varied from 4 to 5 m^3 (1,200 gal).

Initial Gritting Practice

Initially, grit was pumped from each operating softener every eight hours for a period of five minutes. The solids were withdrawn through a 200 mm discharge line using a 227 m^3/h (1,000 gpm) grit pump. This procedure resulted in a daily withdrawal of 57 m^3 or 13 times the volume of the bottom cylinder used for the collection of the grit. If the total of 5,720 kg/d of solids were removed by gritting, the sludge withdrawn would contain roughly10 percent solids.

The *CB&I ClariCone Operating Manual* does not give specific recommendations with regard to gritting possibly owing to different treatment process applications. However, the manual suggests opening the grit drain "for a few minutes each day when the raw water turbidities are high" for those treating river water.

Observation of Gritting

A series of studies was conducted to evaluate the effectiveness of the gritting process and the impact of the established grit removal protocol on the softening process. During a grit removal cycle, samples from a tap in the grit discharge pipe downstream from the grit pump were collected at one-half minute intervals (Fig. 6-2). Each sample was analyzed for percent solids. Thereafter, the supernatant was decanted and the residual solids were dried and weighed.

As shown in Fig. 6-3, after two minutes of grit pump operation, both the solids concentration and weight of grit removed dropped to a steady state. These steady-state concentrations resembled those taken from the bottom of the slurry blanket that had total solids of 1.12%. This suggests that, after two minutes (or 7.6 m^3 volume removed), the gritting process was merely withdrawing slurry from the bottom of the blanket.

Slurry Settling Rates

Based on these results, additional studies were conducted to observe the effect of gritting on the stability of the slurry blanket. Slurry blanket samples from three depths within the softener were collected using the specially adapted sampling apparatus shown in Fig. 6-4.

Figure 6-2 Sample jars containing grit.

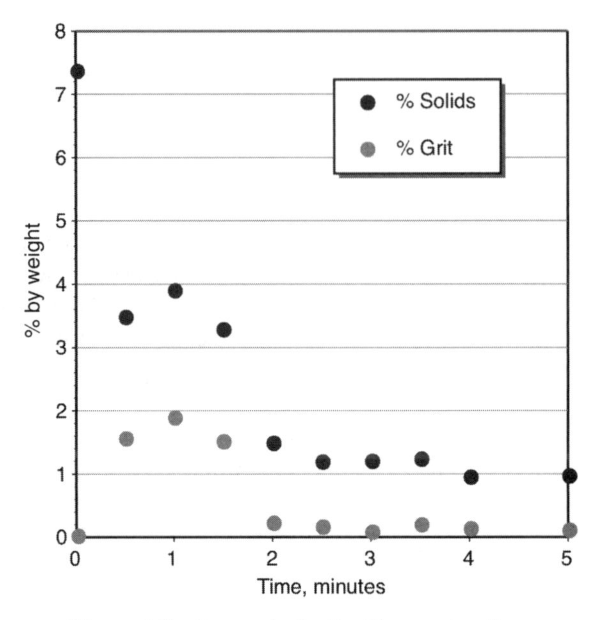

Figure 6-3 Removal of grit with pumping time.

Figure 6-4 Slurry blanket sampling device.

TABLE 6-1 Dry Solids Concentration as a Function of Depth within the Slurry Blanket

Location	Depth	% Solids
Top	0.3 m below top of blanket	0.50
Middle	3 m below top of blanket	0.77
Bottom	5 m below top of blanket	1.12

The percent solids data obtained (Table 6-1) established the baseline for normal solids concentrations within the slurry blanket. In addition, for the slurry from each sampling depth, the gross (surface interface) settling rates were observed (Fig. 6-5). These rates decrease with increasing solids concentration due to hindered settling.

Gritting was then conducted on the softener. Figures 6-6 to 6-8 illustrate the effects that this removal of grit had on the slurry surface settling velocities within the slurry blanket at three discrete times: immediately following gritting, after 30 minutes, and after two hours.

For those samples taken just below the surface of the blanket, the slurry surface settling rates were found to be roughly 9 m/h, both before and after gritting.

At 3 m below the surface, before gritting, the slurry surface settling rate was about 7 m/h. Two hours after gritting, this rate appeared to approach the 9 m/h rate observed from samples taken at the top of the blanket, indicating a less dense slurry blanket.

Finally, at 5 m below the top of the blanket, where the initial solids concentration was 1.12%, the slurry surface settling rate, initially less than 5 m/h, approached 9 m/h two hours after gritting.

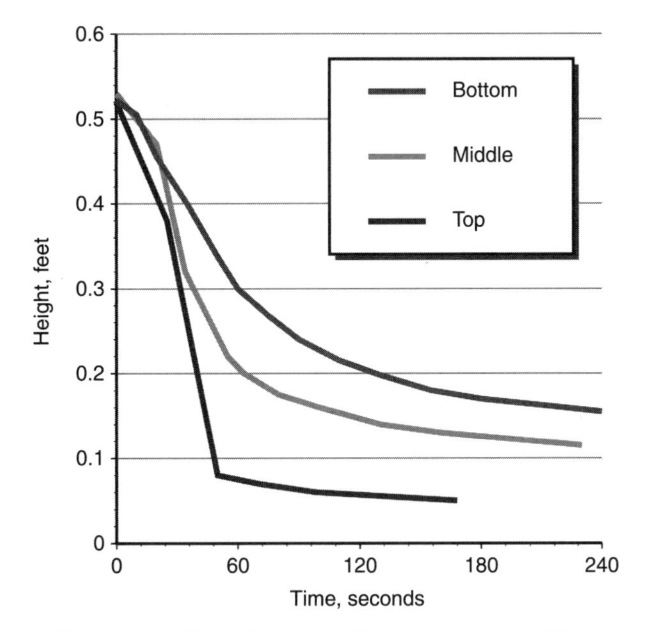

Figure 6-5 Slurry blanket settling rates prior to gritting.

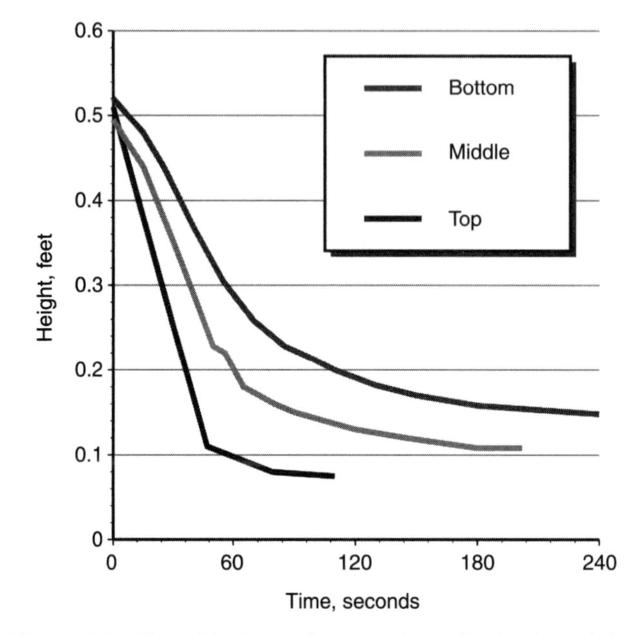

Figure 6-6 Slurry blanket settling rates immediately after gritting.

While the slurry blanket settling rate data may not be accurate or sensitive enough to allow for meaningful conclusions, the general shapes of the grouped settling curves indicate that settling rates at all levels within the slurry blanket converged (increased throughout the depth) following the gritting procedure. This would confirm vertical mixing (and disruption) of the slurry column.

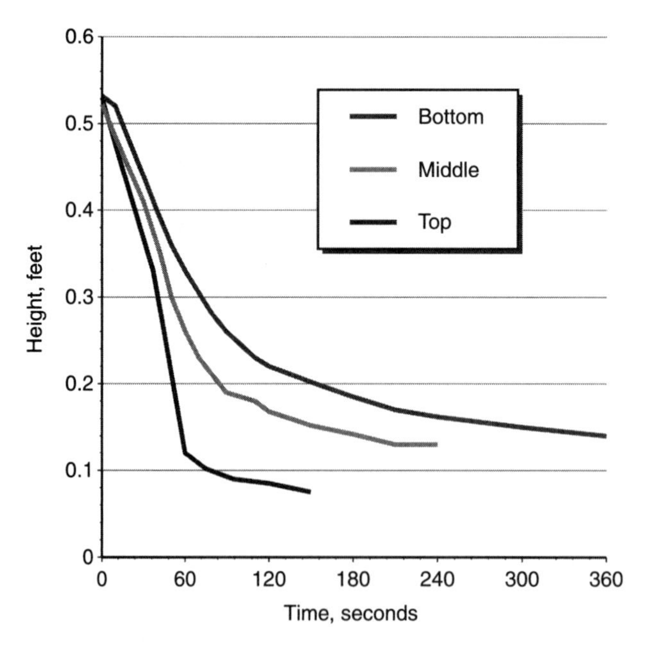

Figure 6-7 Slurry blanket settling rates 30 minutes after gritting.

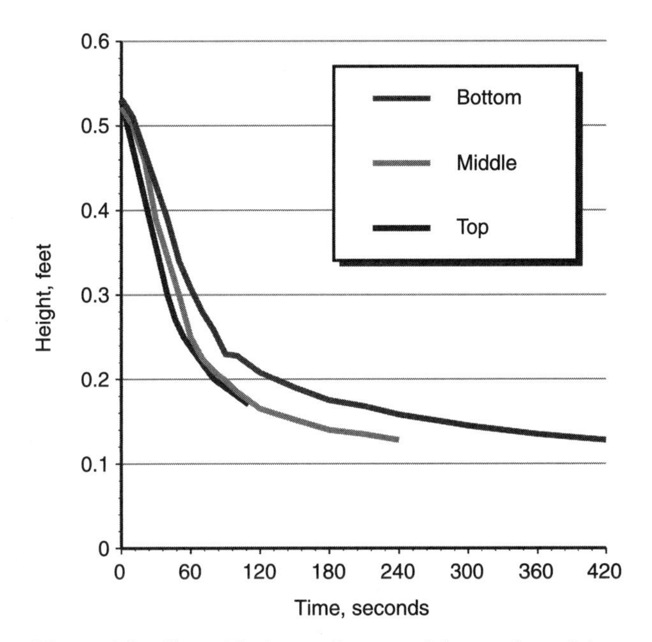

Figure 6-8 Slurry blanket settling rates 2 hours after gritting.

Video

As part of this evaluation, an underwater video camera was used to observe conditions inside the tank. Of particular interest was the video from inside the mixing zone cylinder where lime slurry and influent water enter and grit settles. While this is a region of high turbulence, it appeared that some of the heavier particles of unreacted lime were settling directly into the grit settling compartment, thereby wasting a portion of the applied lime.

Conclusions and Recommendations

The initial gritting study indicated that most of the settled grit at the bottom cylinder of the softener was removed within two minutes. Based on this result, operators were advised to consider incrementally reducing the frequency and duration of gritting towards a target of once daily for a shorter duration that would result in a daily withdrawal of less slurry.

CLARICONE SETTLING VELOCITIES VERSUS APPLIED FLOW

Upward flow velocities in the softener vary with both the applied hydraulic loading and the depth within the inverted cone-shaped tank. Overflow velocities are given in Table 6-2 and are illustrated in Fig. 6-9 for applied hydraulic loadings of 19,000 m^3/d, 22,700 m^3/d, and 26,500 m^3/d (5, 6, and 7 mgd).

At Bloomington, a stable blanket interface appears to form when overflow velocities, v_o, are approximately 5 m/h. If this condition remained constant, the depth from the settled

TABLE 6-2 Overflow Velocities in ClariCone 4 for Hydraulic Loadings at 7, 6, and 5 mgd

Depth, m	Diameter, m	Area, m^2	v_o m/h @ 26,500 m^3/d	v_o m/h @ 22,700 m^3/d	v_o m/h @ 19,000 m^3/d
0	19	290	3.7	3.2	2.7
1.5	16	206	5.1	4.4	3.7
3.0	13	137	7.8	6.6	5.6
4.5	10	82	13.0	11.2	9.2
6.0	8	41	26.4	22.4	18.8
7.5	4	14	77.2	66.2	55.2
9.0	3	9	125.1	107.3	89.2
Bottom	3	9	125.1	107.3	89.2

water surface to the top of the blanket would vary with hydraulic loading and range from 1.2 m to 2.4 m, as shown. Accordingly, the top of the concentrator cone, used to remove (blowdown) the accumulated slurry blanket, will require adjustment each time the hydraulic loading is altered.

In addition, since an operational goal is to maximize collisions and aggregation of floc particles, a deeper, thicker slurry blanket provides more opportunities for collisions and aggregations. The amount of supplemental anionic polymer feed affects the weight of the blanket such that a lower feed should result in a deeper, thicker blanket (lighter particles, slower settling rates) and vice versa. In addition, the energy input (controlled by altering the rotational energy via flows to the inlet) will have an effect on bed depth.

Chemical feed rates to the softeners should be adjusted so that the settling rate of the floc formed at the slurry blanket surface is equal to the rise rate of the water just above the rim of the concentrator cone.

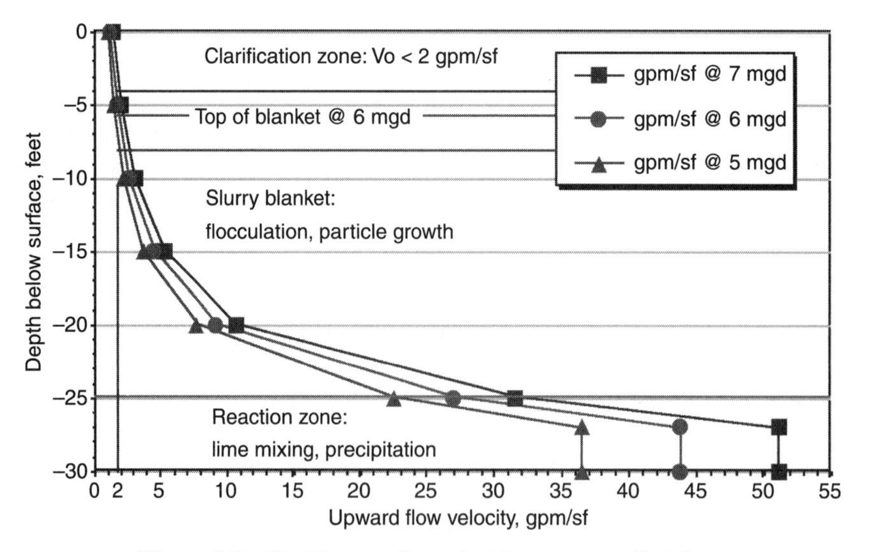

Figure 6-9 ClariCone settling velocities versus applied flow.

GRIT REMOVAL MODIFICATION

As a follow-up to previous tests, solids removals during gritting were observed on ClariCones 3 and 4. These softeners are the first and second to receive slaked lime slurry from the overhead, recirculating, lime delivery system and, owing to differential settling during transport, receive a greater share of the heavy grit than subsequent softening units.

Second to receive lime slurry, ClariCone 4 was gritted using the established protocol in which the settled contents of the cylinder were jetted for five minutes to loosen and resuspend the accumulated grit. After jetting, the grit pumps were then turned on and operated for five minutes until shutdown. As the pumping (grit removal) proceeded, samples of the discharge were taken at one-quarter minute intervals throughout the process.

The samples were analyzed for total dry solids. From Figs. 6-10 and 6-11, it is evident that the highest solids content was found within one-half minute into the pumping cycle. After one minute, the percent solids removed during the gritting procedure dropped to levels approximating those found in bottom of the slurry blanket (~1%).

The controlled grit test of ClariCone 4 confirmed that this gritting procedure disrupted the slurry column, created turbidity in the clarification zone, and resulted in a lime overfeed.

In an effort to moderate this disruption, a modified procedure was attempted in gritting Cone 3. For this procedure, the grit pump was first activated for two minutes without jetting.

Figure 6-10 Grit removed over 5 minute gritting period.

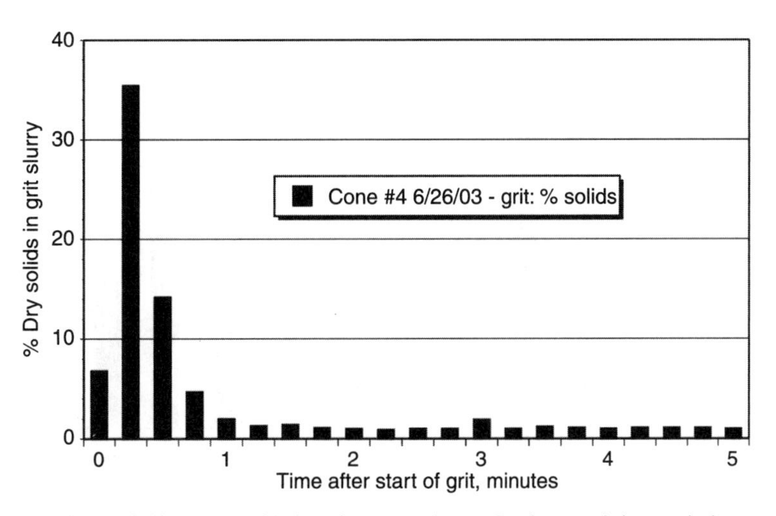

Figure 6-11 Dry solids in grit removed over 5 minute gritting period.

Figure 6-12 Grit removed during revised gritting procedure.

While pumping continued, the jets were then opened for two minutes and turned off. Finally, the pump was run for one minute following closure of the jets. Partly because Cone 3 had not been gritted in some time, a high concentration of grit was initially removed by this procedure. Then, as the jets were turned on, additional grit was suspended in the cylinder and removed by the grit pump. Overall, very substantial amounts of solids were removed over the five minute gritting period.

Figures 6-12 and 6-13 illustrate the removal of solids during the modified gritting process. The bars illustrate the peak percent solids removed within the first minute and, again, after jetting.

Figure 6-14 shows the fraction of the solids removed that were acid insoluble (true grit).

According to the manual of operation provided for the ClariCone, water jets are used for start-up after long inactive periods (>16 hours). Thereafter, the jets are to be turned on for 5 to 10 minutes to agitate the accumulated solids before introducing the raw water flow. The jets may also be used (for a few minutes) to suspend grit and heavy solids before wasting from the grit drain. Bloomington staff has modified (reoriented) these water jets to point *downward* to more effectively aid in the removal of the compacted solids.

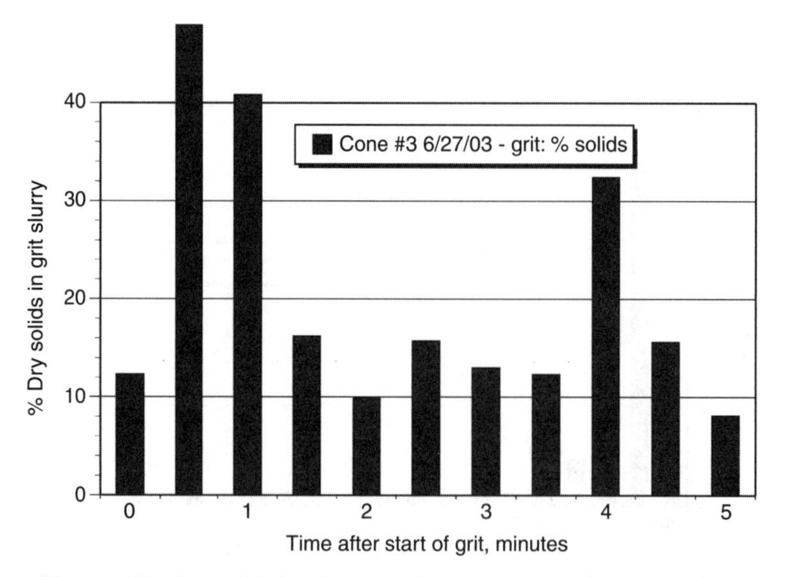

Figure 6-13 Dry solids in grit removed during revised gritting procedure.

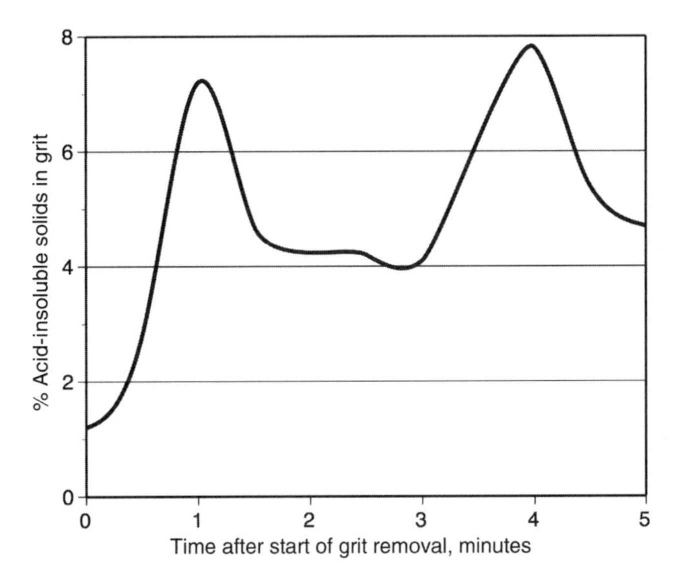

Figure 6-14 Acid insoluble fraction removed during revised gritting procedure.

RECARBONATION FOLLOWING SLURRY BLANKET UPSET

Observations of the effect of grit removal practice on slurry blanket stability has indicated repeatedly that the procedure results in upsets throughout the depth of the slurry blanket column. As this also leads to unanticipated, increased quantities of resuspended lime in the blanket, the softener effluent pH is often observed to increase to well over 11.

Upon the subsequent addition of carbon dioxide in the recarbonation basin, post-precipitation of calcium carbonate turns the recarbonated water turbid (Fig. 6-15).

Figure 6-15 Post-precipitation of calcium carbonate in recarbonation basin.

Figure 6-16 Recarbonation basin under normal operating conditions.

Alternately, when the slurry blanket is stable and effluent pH is between 10 and 11, the water in the recarbonation basin may be so clear that the bottom of the recarbonation basin is visible (Fig. 6-16).

Lime overfeed from gritting Cone 4 two hours earlier was also reflected in a large increase in acid-soluble turbidity when carbon dioxide was subsequently applied in the recarbonation basins (Fig. 6-17).

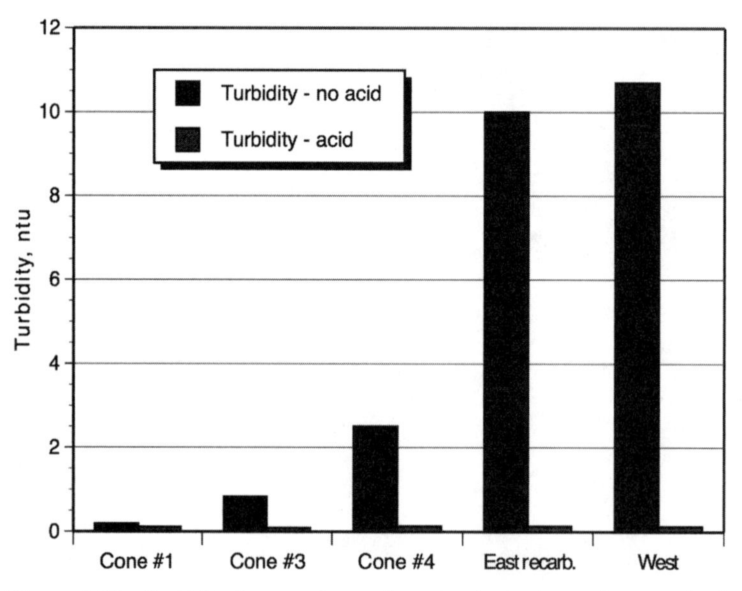

Figure 6-17 Turbidity increase in recarbonation basin due to lime overfeed.

PARTICLE SIZE ANALYSIS: EFFECT OF GRITTING

Samples were taken of the slurry blanket of ClariCone 4 to obtain particle size profiles as a function of depth, before and after gritting. With the exception of the two samples taken from the clarification zone, all slurry blanket samples were allowed to settle to remove most of the blanket slurry. Only the supernatant was taken for particle analysis using the electronic particle counter. A second series of these samples was taken for particle counting after the samples had been *acidified* to dissolve calcium carbonate and magnesium hydroxide precipitates. The results of both sampling series are shown in Fig. 6-18.

Before gritting, most of the particles in the clarification zone samples (surface, 0 m) were found to be very small ($<8\ \mu m$). Alternately, most of the particles in the slurry blanket supernatant were larger than 8 μm. These results confirm the expectation that almost all particles larger than 8 μm are removed within the slurry blanket.

Acidification of all samples resulted in significant reductions in both particle sizes and numbers. While large particles were virtually absent in the acidified samples, total particle numbers declined by about 80%.

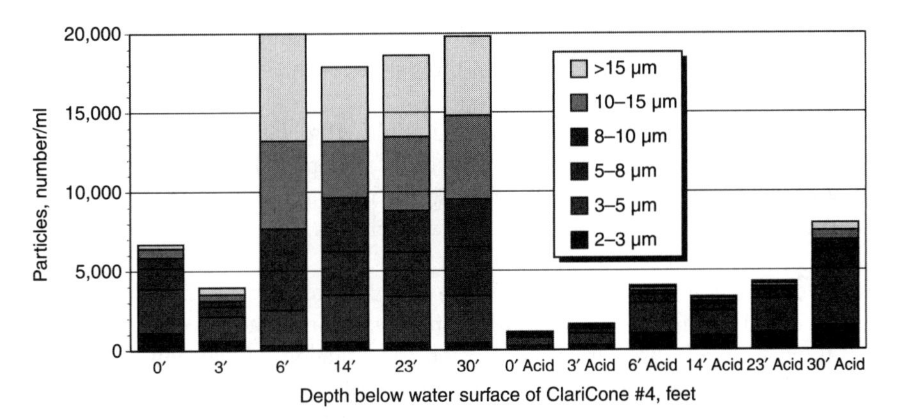

Figure 6-18 Particle count distributions with depth in softener, without and with acidification.

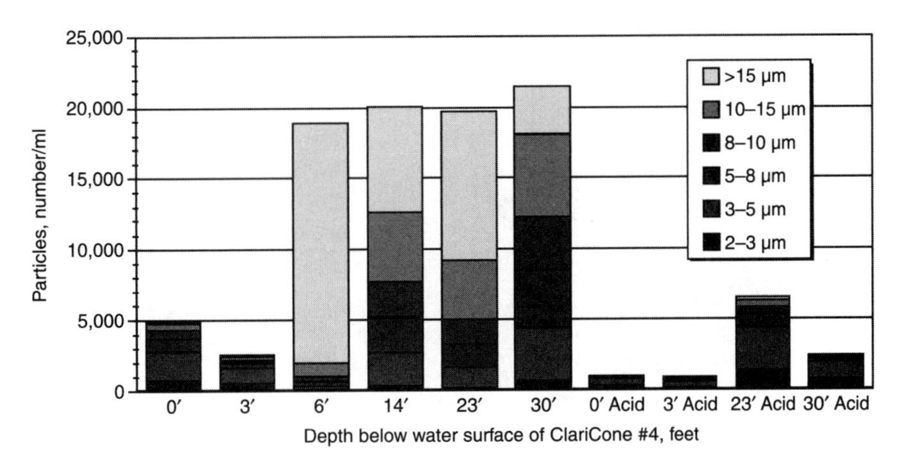

Figure 6-19 Particle count distributions with depth in softener, before gritting.

The majority of particles in the acidified samples were in the 2 to 5 μm size range. For the same equivalent mass, particles in this size range contribute far more to the measurement of turbidity (light scattering) than larger particles.

This particle profile analysis procedure was replicated just prior to gritting ClariCone 4. The results (Fig. 6-19) showed that the numbers of particles were very similar to those recorded in the first trial. However, the particles in the slurry blanket supernatant tended to be larger. Once more, acidification resulted in major reductions in particle sizes and numbers.

POST GRIT

After gritting, significant changes in particle numbers and distribution were evident (Fig. 6-20). The numbers of particles found in the clarification zone tripled or quadrupled. Moreover, the sizes of these particles were far larger, indicating reduced settling efficiency. Following gritting, the clarified water more closely resembles the particle-rich supernatant from the slurry blanket.

While acidification reduced the numbers of larger particles, it reduced total particle numbers by only 50% to 75%. This particle count data confirmed visual observations indicating increased turbidity and instability of the surface of the slurry blanket following the gritting procedure.

A plot of all total particle count data versus turbidity (Fig. 6-21, top curve) shows a nonlinear relationship. A similar relationship is evident for all particles larger than 8 μm versus turbidity (middle curve). However, the very strong relationship between particles smaller than 8 μm and turbidity is only evident at low turbidities (<1 ntu) where small particles accounted for most of the measured turbidity.

Estimated retention times in various ClariCone zones are shown in Table 6-3 for various flow rates. Based on analyses for calcium and magnesium ion, it appears that the dissolution of the influent lime slurry followed by the precipitation of calcium carbonate and magnesium hydroxide occur rapidly, within minutes. Mixing in the slurry blanket provides tapered (reduced energy) flocculation as upward flow velocities decrease.

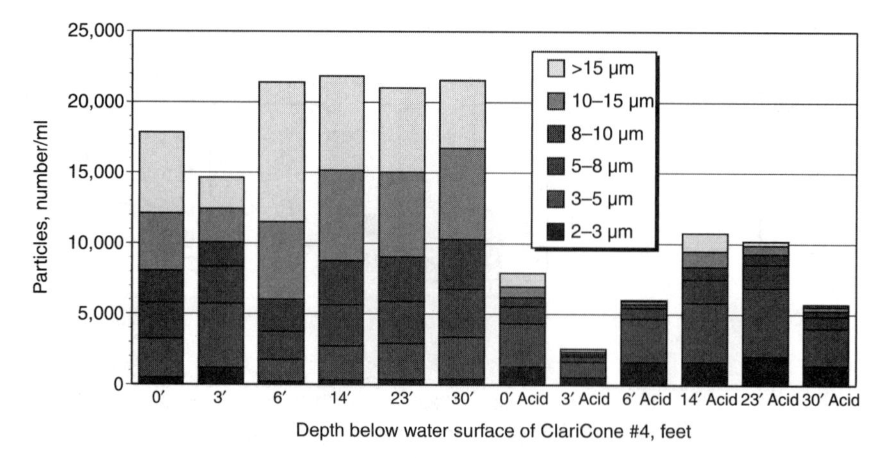

Figure 6-20 Particle count distributions with depth in softener, following gritting.

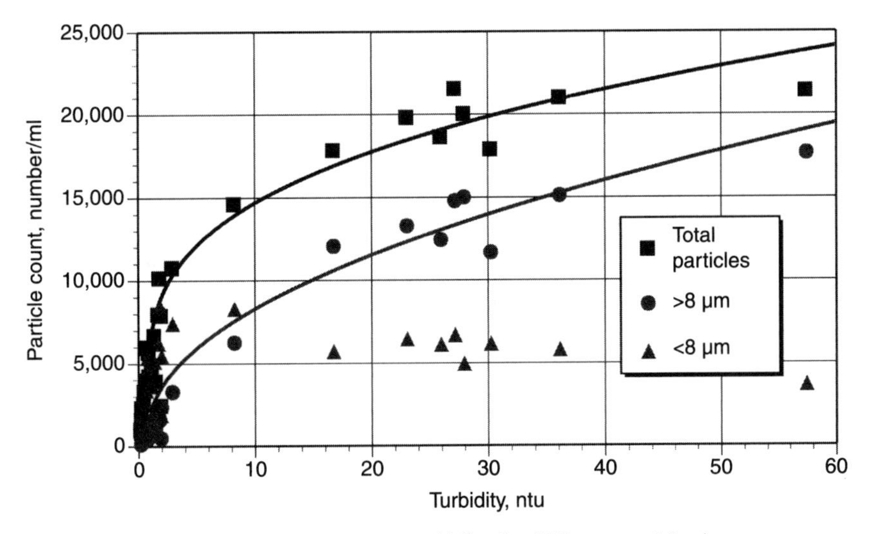

Figure 6-21 Particle count versus turbidity for different particle size groups.

TABLE 6-3 Retention Times in ClariCone 4 for Various Hydraulic Loadings

Zone	Volume, m^3	Retention, min @ 26,500 m^3/d (7 mgd)	Retention, min @ 22,700 m^3/d (6 mgd)	Retention, min @ 19,000 m^3/d (5 mgd)
Clarification	390	21	25	30
Upper blanket	270	15	17	20
Middle blanket	172	9	11	13
Lower blanket	96	5	6	7
Flocculation	43	2	3	3
Reaction	18	1	1	1
Mixing	20	1	1	2
Total volume, m^3	725	55	64	77
Clarification zone when	Concentrator @ 1.5 m	21	25	30
Total—clarification zone	= Slurry blanket	34	39	47

While economic in terms of softening, the short retention times in the softening units may contribute to the adverse impact of prolonged gritting on the hydraulic profile of the slurry blanket. Along with prolonged gritting, introduction of unreacted lime into the softening units during jetting, explained the adverse impact to the pH and hydraulic profiles of the slurry blanket.

pH AND CONDUCTIVITY

The disruption of the ClariCone 4 slurry blanket following gritting is strongly indicated by the measurement of pH. Figure 6-22 illustrates a pH profile through ClariCone 4 that shows the sharp increase in pH resulting from lime feed at depths greater than 9 m. Two

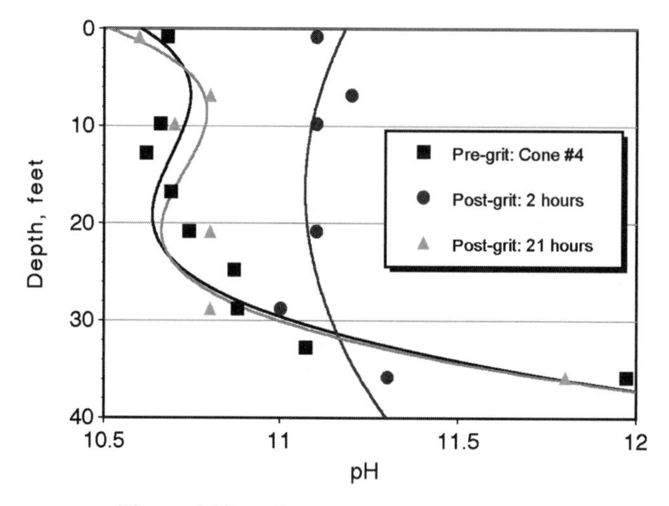

Figure 6-22 Effect of gritting on pH profile.

hours after gritting, this profile has changed and pH is more nearly uniform throughout the depth of the softener. Finally, 21 hours after gritting, the pH profile is restored to its original shape. It may be also inferred from the stable profiles that lime dissolution and softening are near complete at depths above 6 m.

Comparative pH profiles for all four ClariCones are shown in Fig. 6-23. At this time, maintenance personnel were making modifications to improve the performance of ClariCone 2. This is reflected in a high surface water pH and a profile that indicates a pH inversion in that unit.

The conductivity profiles for ClariCone 3 (Fig. 6-24) indicate a stable, vertical profile down to 9 m. However, 10 minutes after gritting, this profile is distorted, but appears to recover in one hour.

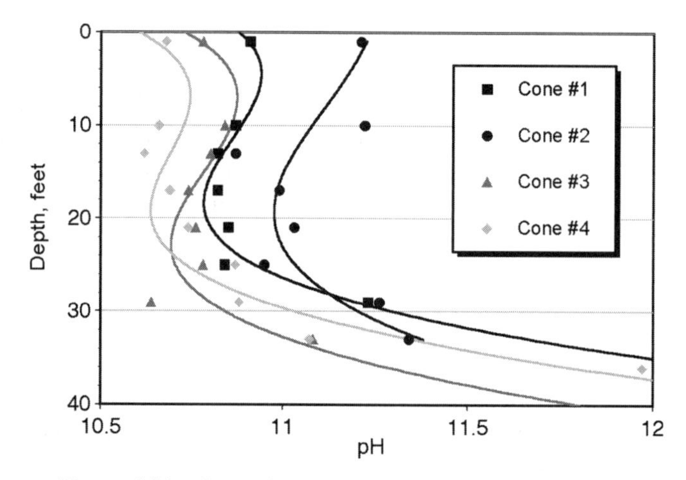

Figure 6-23 Comparison of pH profiles in four ClariCones.

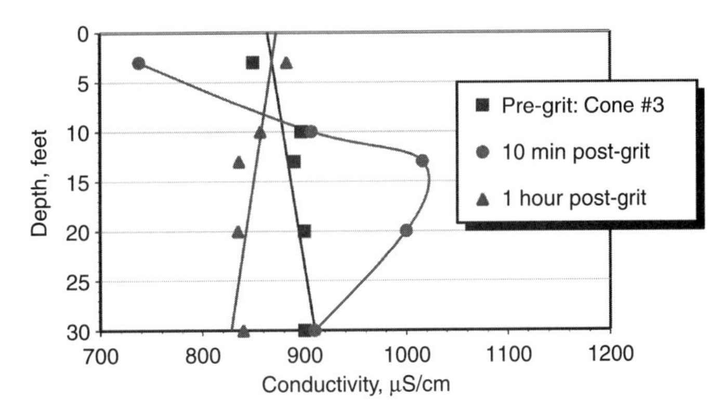

Figure 6-24 Effect of gritting on conductivity profile.

EFFECT OF GRIT REMOVAL ON BLANKET STABILITY

The pH profile obtained in ClariCone 4 before gritting varied from 10.7 at the surface to approximately 12.0 at the bottom of the cone. The high pH at the bottom reflected the influx of the lime slurry. The resulting slurry blanket formed during normal softening (production cycle) was stable, with a clear, smooth interface and a very low effluent turbidity. The pH decreased as precipitation took place and alkalinity was removed from solution.

From the rapid decrease in pH, it appeared that the lime suspension went into solution and hardness precipitation occurred within minutes in the lower section of the softener. Thereafter, the slurry blanket provided 20 to 30 minutes of additional contact with previously precipitated particles for flocculation and particle growth.

After the grit settling compartment had been jetted for five minutes and the grit pumps turned on to remove the dislodged solids, the appearance of the surface of the slurry blanket changed. The surface was obscured by solids rising from the blanket and the clarification zone became noticeably turbid. Large quantities of entrained unreacted lime were resuspended from the sludge storage compartment during jetting.

Two hours after the gritting, the vertical pH profile through the ClariCone had changed significantly. The pH at the surface had risen to 11.1 and the entire column was comparatively uniform in pH, averaging 11.2. At this pH, virtually all the magnesium is precipitated as magnesium hydroxide and the effluent water is, instead, high in calcium ion. In effect, the softener effluent becomes a dilute solution of *lime water*, calcium hydroxide. The subsequent addition of carbon dioxide during recarbonation, therefore, forms a calcium carbonate precipitate.

Twenty hours following the gritting procedure, a stable blanket was again observed and a vertical profile confirmed establishment of a pH gradient similar to the one that existed before gritting occurred.

REVISED GRIT REMOVAL PROTOCOL

A subsequent study was conducted using ClariCone 3 to determine whether a modified gritting procedure would eliminate the upset of the slurry blanket and avoid the problems created by lime overfeed. The revised protocol called for the grit pump to be turned on

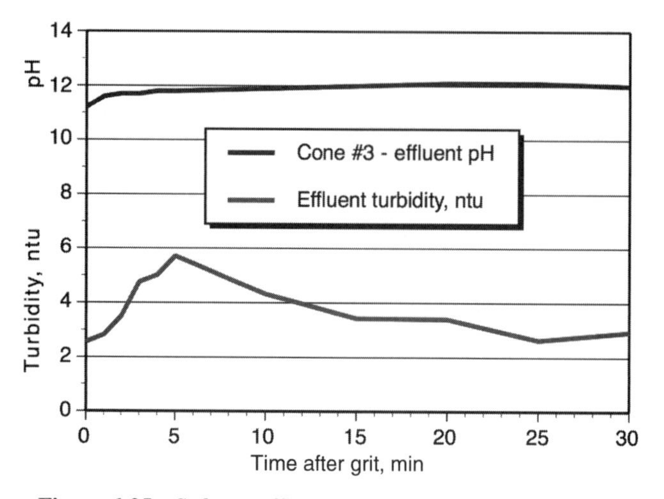

Figure 6-25 Softener effluent pH and turbidity after gritting.

for two minutes for solids removal. While the pump continued running, the jets were activated for two minutes then turned off. The grit pump was allowed to run for another minute to remove the solids dislodged by the jetting. In total, this procedure took five minutes.

Vertical pH profiles in Cone 3, before and after gritting, indicated that the slurry blanket was still destabilized by the modified gritting procedure (Fig. 6-25). The stable pre-grit blanket exhibited a lower pH (10.7) at the top of the bed than at the base (pH 11.6), which received the introduction of the lime feed.

Ten minutes after gritting, the pH was found to increase markedly in the middle of the slurry blanket. This was found to be due to entrained calcium hydroxide from the sludge storage compartment that was forced upward by the turbulence created during gritting and jetting. Atypically low pH (10.0) at the bottom of the ClariCone was attributed to removal of influent lime feed by the grit pump.

After one hour, the entire slurry blanket appears to have reversed its initial stratification and effluent pH values exceeded 11.0. At these high pH values, all the magnesium in the influent water is precipitated and calcium ion has increased due to the solution of the lime feed.

Following a lime overfeed or slurry blanket upset, post-precipitation of calcium carbonate upon recarbonation is certain. As a result, turbidity not only increases markedly in the recarbonation basin, but unsettled calcium carbonate reaches the filters, shortening filter runs. Correcting this problem atop the filters sometimes requires the supplemental short-term addition of coagulant or polymer atop the filters to avoid high finished water turbidities.

Correction of the slurry blanket upsets caused by gritting has the potential for saving significant chemical and operational costs. Most obvious is the saving of carbon dioxide. If the pH of the softener effluents can be maintained in the range of 10.5 to 10.8 throughout the year, less carbon dioxide will be consumed and less sludge produced. Lower turbidities reaching the filter will reduce the frequency of polymer use immediately before filtration and, operationally, reduce the amount of filtered water required for backwash.

Calcium carbonate *accretions* are formed as part of the grit removed from the softeners. This material is comparatively heavy and dense as compared with calcium carbonate

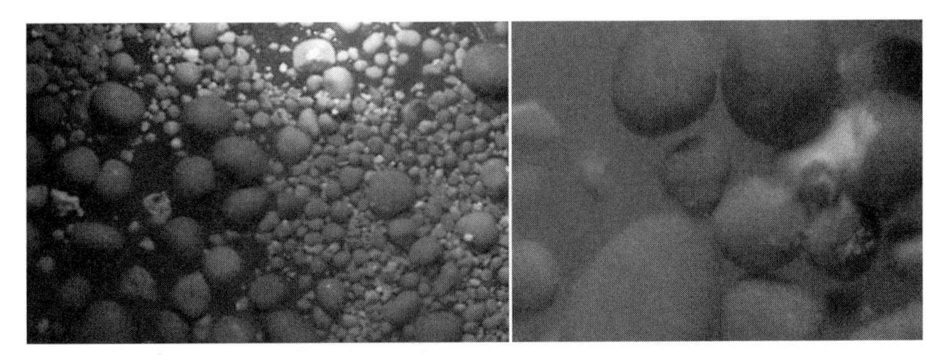

Figure 6-26 Calcium carbonate accretions.

Figure 6-27 Calcium carbonate accretions on sand grains.

sludge. The accretions are believed to be formed around the nuclei of small inorganic particles making up part of the lime slurry feed. Due to a plugged grit line, grit was not removed from Cone 3 for over a month, allowing time for the large, dense particles shown in Fig. 6-26 to form.

In another lime softening process (in the Permutit Spiractor reactor), this characteristic of calcium carbonate precipitation is taken advantage of to produce a solid residue that is dense and readily dewatered. To facilitate the formation of the accretions, sand is added to the top of the Spiractor to provide the nuclei on which precipitation will take place. The solids formed are discharged from the bottom of an elevated hopper.

The material shown in Fig. 6-27 represents the dry residue (solids) from the lime softening of 0.2 m^3 of water from a hardness of 215 to 35 g/m^3 as calcium carbonate equivalent. The solids are the equivalent of about 1 liter of 3% lime slurry. The sample was distributed by Permutit (Sybron Corporation) from the Spiractor installed at the Columbus, Nebraska Water Treatment Plant.

However, in the bottom cylinder of the ClariCone and in the sludge holding tank at Bloomington, these accretions create an operational problem as the particles are not readily resuspended and removed from the settling compartment. The plugging problem was subsequently solved by replacing the existing grit pump with a tri-lobed rotary slurry pump. The rotary pump was able to handle the grit load much more reliably than the original. The problem with post-gritting pH upsets was solved by raising the location of the lime feed discharge point in the clarifiers, so that less unreacted lime would be entrained with heavy grit into the sludge storage compartment.

7

LIME SOFTENER PERFORMANCE ENHANCEMENTS

MODIFIED LIME SLURRY FEED POINTS

Continuing studies were conducted to evaluate and enhance the operational performance of the lime softening process. Proposed modifications were directed at:

- achieving more effective utilization of lime,
- minimizing slurry blanket upset during gritting,
- avoiding transient lime overfeeds, and
- reducing subsequent dose requirements for carbon dioxide.

The previous evaluations of Bloomington's ClariCones had shown that a portion of the influent lime slurry, when injected into the port at the bottom of each unit, settled directly into the grit storage compartment (bottom cylinder). This occurred because the larger particles in the lime slurry had a downward settling velocity that exceeded the upward flow velocity of the influent mixture in the bottom cylinder. As a result, this heavier fraction of unreacted lime was subsequently withdrawn along with grit during each grit wasting cycle.

Based on these observations, the lime feed points to all four ClariCones were significantly modified to permit lime to be added within the slurry blanket at selected elevations *above* the bottom cylinder (Fig. 7-1). Lime feed discharge at this higher elevation in the softener allowed the lime suspension to mix and remain in contact with the influent water and slurry blanket for a longer period before heavier components settled into the bottom cylinder. This more flexible feed system also allowed for vertical adjustment of the lime slurry feed point within the slurry blanket. The lime slurry flow is controlled using an adjustable pinch valve (Fig. 7-2) while a flow meter sends flow rate data to the flow controller and the SCADA system (Fig. 7-3).

Water Treatment Plant Performance Evaluations and Operations. By John T. O'Connor, Tom O'Connor, and Rick Twait

Figure 7-1 Modified lime slurry feed system (flexible hoses from overhead recirculating slurry line).

Figure 7-2 Adjustable pinch valve controls lime slurry flow.

Figure 7-3 Lime slurry flow meter sends rate data to flow controller.

EVALUATION OF GRIT REMOVAL PROCESS

The gritting process had previously been evaluated to determine whether adverse effects on the slurry blanket stability were associated with pregrit water jetting to resuspend grit. Tests of reduced gritting time as well as substitution of within-cycle jetting were conducted to observe their effects on reducing hydraulic transients (Chapter 6).

Modification of the gritting procedure was undertaken to save lime by moderating periodic lime overfeed. As indicated by transient increases in pH of the softened and settled water, lime overfeed is observed primarily during periods when gritting takes place. This is also indicated by short-circuiting of flow within the slurry blanket leading to turbulence (Fig. 7-4) and upwellings (Fig. 7-5) at the blanket surface. Since the lime slurry influent

Figure 7-4 Surface upset of slurry blanket following gritting.

Figure 7-5 Upwelling in slurry blanket.

feed continues uniformly, even while a portion of the softener influent water is withdrawn from the drain as part of the gritting process, the effective lime softening dosage increases. Also, since entrained, unreacted lime was introduced into the clarifier from the sludge storage compartment during the water jetting phase of the gritting process, the effective lime softening dosage increases.

When lime overfeed occurs, there is an increase in the consumption of carbon dioxide in the recarbonation basin. The reaction between the lime and carbon dioxide not only results in increased calcium carbonate post-precipitation (turbidity), but also results in increased solids loading on the filters (Fig. 7-6). Where post-precipitation is incomplete in the recarbonation basin, turbidity increases may also be observed post-filtration.

Based on the earlier study results, each softener is now gritted once every 24 hours. To resuspend the heavier particles in the grit, water jets are activated for 2.5 minutes before

Figure 7-6 Lime overfeed resulting in calcium carbonate post-precipitation.

Figure 7-7 Concentrator cone within dewatered ClariCone.

Figure 7-8 ClariCone effluent trough.

opening of the waste line and activation of the grit pump. Thereafter, the grit pump was run for 7 minutes at a rate of 4 m^3/min.

Alternately, the *blowdown* of the slurry blanket from the concentrator cone (Fig. 7-7) within the ClariCone is programmed for 15 minutes every four hours. Finally, the settled effluents from the softeners are collected in horizontal troughs that span the diameter of the surface of the ClariCone (Fig. 7-8).

REDUCTION IN BLANKET UPSETS, LIME OVERFEEDS, AND CARBON DIOXIDE REQUIREMENTS

Previous evaluations of softener performance documented the adverse effect of the automated gritting process on the stability of the slurry blanket. The introduction of water *jetting* to fluidize the compacted grit, followed by the removal of the grit suspension by pumping at a high rate (4 m^3/min), created transient vertical velocities that appeared to destabilize the blanket. This was evidenced by marked distortions in the softener's vertical pH and particle size profiles. Following each gritting, high pH water was discharged, thereby necessitating increased amounts of carbon dioxide for stabilization.

Bloomington plant operators commonly observe the effects of upsets by observing both slurry blanket surface upwellings and turbidity increases in the recarbonation basin. It has been determined that the carryover of high pH softener effluent (due to lime overfeed) into the recarbonation basin results in the post-precipitation of calcium carbonate when carbon dioxide is applied. This not only turns the contents of the recarbonation basin (blue-green) cloudy, but also increases the solids loading applied to the plant's filters.

The effects of lime overfeed and post-precipitation during recarbonation are not limited to heavier filter loadings and an increased possibility of turbidity exceedances. With post-precipitation in progress, there is also a build-up of calcium carbonate on the sand filter media underlying the granular activated carbon filter cap.

RECARBONATION BASINS: POST-PRECIPITATION PARTICLE SIZE ANALYSIS

A particle size analysis was performed on samples taken from the east and west recarbonation basins using Bloomington's electronic particle counter. The total numbers of particles enumerated in each basin were modest and comparable (Fig. 7-9). However, the distribution of particle sizes indicated that most of the particles present were larger than 8 μm. This large size would indicate that numerous smaller particles (incipient precipitates of calcium carbonate) had already grown rapidly to a size that would be readily filterable by Bloomington's dual media filters.

However, if calcium carbonate post-precipitation continued to occur within the filter beds, turbidity exceedances might be detected in the finished water. Where turbidity exceedances have been detected, it is believed that post-precipitation of calcium carbonate has been the principal cause.

Although the recarbonation basins are decades old and occupy a large surface area (Fig. 7-10), they appear to remain effective in absorbing the bulk of the applied carbon dioxide. Tests of the atmosphere above the surface of the basins at the point of carbon dioxide application did not reveal elevated levels of the gas. These measurements, using

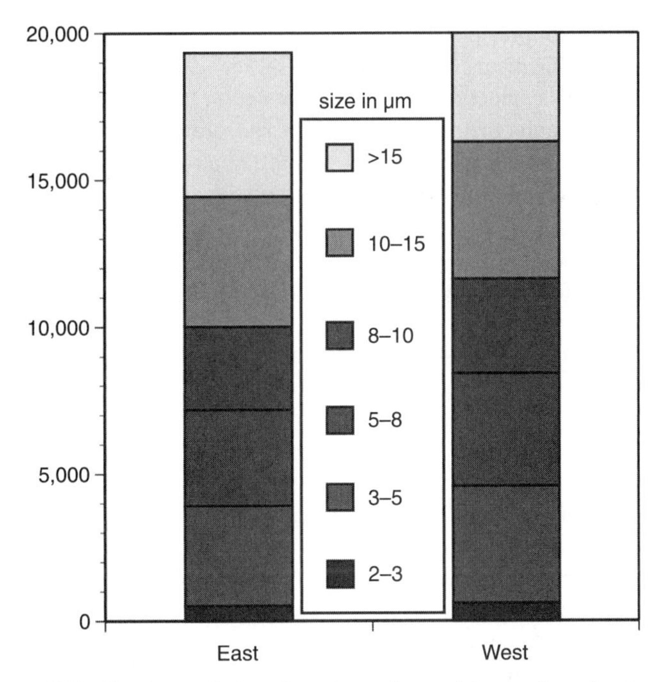

Figure 7-9 Number and size of particles observed in recarbonation basins.

a CO_2/temperature monitor, indicate that substantial amounts of the gas are not escaping to the atmosphere.

The periodic post-precipitation of calcium carbonate is evident from the milky-white appearance of the recarbonated water that obscures the concrete baffle walls. In addition, post-precipitation creates several maintenance problems. One is evident from the encrustation of calcium carbonate on all the effluent launders (background). The other is the

Figure 7-10 Recarbonation basin showing serpentine effluent launders (background) and baffle wall.

resultant accumulation of precipitate in the bottom of the basin. Routinely, because there are no sludge removal mechanisms, the basins must be manually dewatered and cleaned.

Plans for future replacement of the two recarbonation basins call for smaller and more operationally efficient units that will be enclosed and protected from the influx of leaves and windborne debris. It is anticipated that both carbon dioxide feed requirements and operational costs will be reduced by this new installation.

RESULTS OF THE EVALUATION OF THE GRITTING PROCESS

Tests of the gritting process indicate that reductions in pregrit jetting, changing the location of the lime feed point, and the length of the grit removal cycle may result in far less disturbance of the slurry blanket column.

Figure 7-11 shows that, before gritting, the pH in the upper portion of the slurry blanket and clarification zone was near constant at about 11.2. Following the grit (10 minutes; 1 hour), the pH in this column increased to approximately 11.3. After 2 hours, the pH again approached the initial pH of 11.2. This degree of pH disturbance would be considered minimal in a lime softening reactor.

While further operational changes in the gritting process may still be desirable, the current procedure, which eliminates the previously documented excursions in effluent pH, is considered near optimal.

The influence of the lime slurry feed blending with the pH 8.4 influent water can be most readily observed at the 7.5 m (25 ft.) depth where pH has increased and varies widely. This is in the region in which both lime dissolution and the softening reactions occur. From the near-steady-state pH observed above the 6 m (20 ft.) level, it appears that the softening reactions are complete and that only particle growth (flocculation) is occurring within the slurry

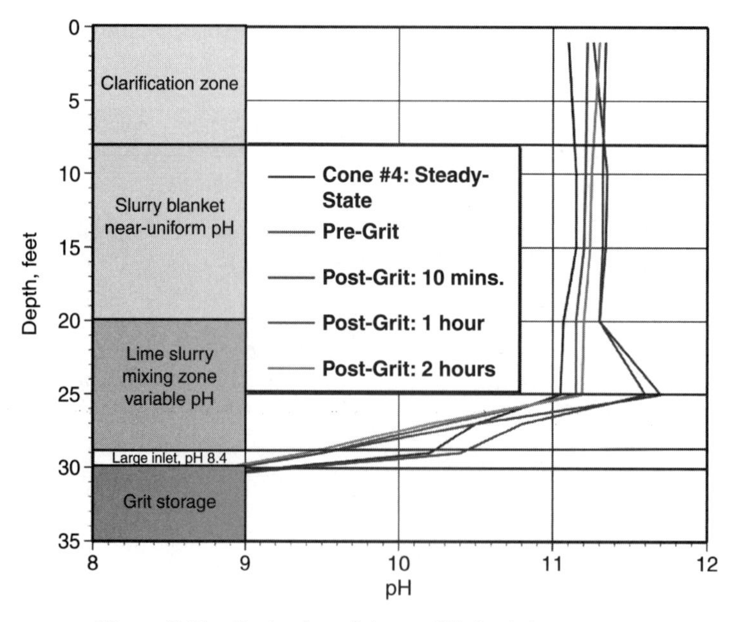

Figure 7-11 Evaluation of the modified gritting process.

column. When the system is stable (no transient vertical flow), there is a clear line of demarcation between the surface of the slurry blanket and the clarification zone.

EVALUATION OF THE INLINE DEGRITTER

Bloomington's automated batch lime slakers have reduced many operational problems and resulted in a more uniform lime slurry feed to each of the four softening units. However, the

Figure 7-12 Grit removal apparatus.

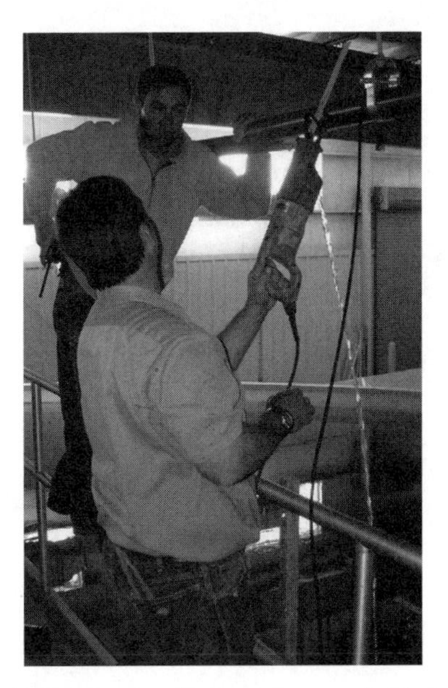

Figure 7-13 Sawing the lime slurry recirculation line.

Figure 7-14 Installing the apparatus.

Figure 7-15 Horizontal PVC grit settling compartment.

Figure 7-16 Monty llama with air-activated valve.

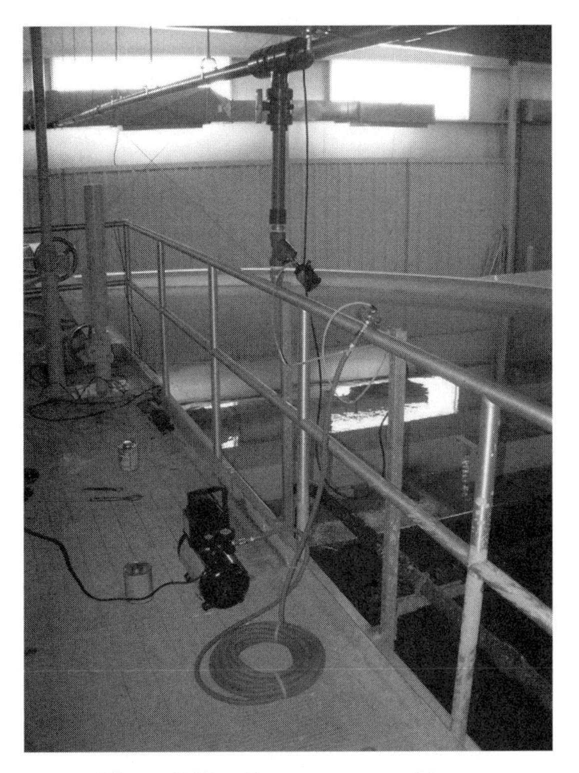

Figure 7-17 Air compressor and hose.

batch slaking system suffers from the fact that the grit entrained in the raw lime cannot be removed during the slaking process. As a result, slaked lime plus entrained grit is transferred to the lime slurry storage facility.

From storage, the lime slurry is constantly recirculated through a flexible pipe loop positioned above the softeners. This procedure minimizes settling of lime solids and helps to maintain uniform slurry density and quality. However, the entrained grit, which is also continually recirculated, causes increased abrasion and wear of the piping and storage facility.

The original concept of this lime slurry feed procedure was that the required quantity of lime for each softening unit would be withdrawn from the overhead recirculating loop and injected into a lime slurry feed port at the base of each softener. Here the slurry would blend with the coagulated influent water to dissolve and initiate the softening reactions. While this objective was readily accomplished with respect to the lime feed, it turned out that the grit was not being distributed uniformly among the softening units. The first unit to receive lime slurry from the recirculating loop received a disproportionately large portion of the grit. This heavy material rapidly settled into the bottom cylinder of the softener, impeding the removal of solids from the bottom cylinder in the gritting process.

To mitigate the problem of the abrasive and disproportionate distribution of the grit in the influent lime slurry, Bloomington plant maintenance personnel developed a unique assemblage to intercept a portion of the grit traveling through the overhead slurry line. The device provided for settling of a portion of the slurry in an expanded (reduced velocity) section of the recirculating slurry line. From this section, every few minutes, an air-activated valve opened, allowing a portion of the accumulated solids, primarily grit, to be discharged to the surface of the softener. The installation of the grit removal apparatus is illustrated in the series of Figs. 7-12 to 7-17. The grit clearing problem was eventually resolved by initiating a clarifier lime feed valve flushing routine. The routine is controlled by the SCADA system, and can be modified by the plant operators.

8

LIME SOFTENER OPERATIONAL ENHANCEMENTS

While the ClariCone softener is an advancement in lime softening technology in that it eliminates most of the mechanical complexity of previous generations of upflow lime contact clarifiers (rotating sludge rakes, shrouds, mixing/blending compartments, radial and peripheral weirs), it creates new operational problems if its complex hydraulics are not properly adjusted. The current evaluation was directed at observing and refining the operational performance of the ClariCone softeners at the Bloomington Water Plant.

As indicated in the CB&I *ClariCone Operating Manual* (March 1999), these units require the establishment of the following operator-controlled variables:

- treatment chemistry (e.g., lime, ferric coagulant, anionic polymer additions; pH control),
- influent flow rates though the large (0.76 m) and small (0.3 m) tangential inlets entering the base of the ClariCone,
- inlet flow and velocity (energy input) control for mixing through the use of butterfly valves,
- maintenance of maximum slurry blanket height without solids loss to effluent trough,
- adjustment of the height of the concentrator cone for concentrated solids removal,
- adjustment of the slope of the effluent channel trough to minimize carryover of solids,
- determination of the frequency and duration of (1) grit removal and (2) blowdown.

Water Treatment Plant Performance Evaluations and Operations. By John T. O'Connor, Tom O'Connor, and Rick Twait

TREATMENT CHEMISTRY

The chemistry of the plant process is determined by the source water quality and treatment goals. Bloomington requires the precipitation of calcium carbonate and magnesium hydroxide through lime addition to achieve hardness reduction. In addition, coagulation with ferric sulfate, assisted with cationic polymer, is used to entrain and remove particles and organisms from the lake water sources. Chemical dosages are adjusted to match changes in influent and effluent water quality, as well as to compensate for significant seasonal water temperature changes.

A substantial quantity of solids is produced in lime softening. The precipitates formed create a dense slurry blanket in the softener. This blanket reputedly serves as a contact medium for the more complete precipitation of hardness and for entrainment of particles present in the lake waters. Anionic polymer applied to the softener influent is used to aid in flocculation and limit the carryover of lighter particles.

At steady state, the lime softening process produces a well-defined slurry blanket surface that is largely composed of discrete, visible flocs that settle at velocities around 0.1 m/s. These settling velocities are far greater than the maximum overflow rate, 0.05 m/s, at the level of the effluent trough. As a result, the turbidity of the clarifier effluent is consistently low ($\approx$1 ntu) and is principally due to the loss of precipitates created by lime softening. Such low turbidities would indicate excellent clarifier performance.

FLOW RATES TO CLARICONES

By design, the inflows to each softening unit should result in surface overflow rates ($V_0 = Q/A_s$) that fall in the range of 1 to 5 m/h. Accordingly, Bloomington's larger (20 m diameter) ClariCones, which have a maximum surface area (A_s) of 290 m^2, are

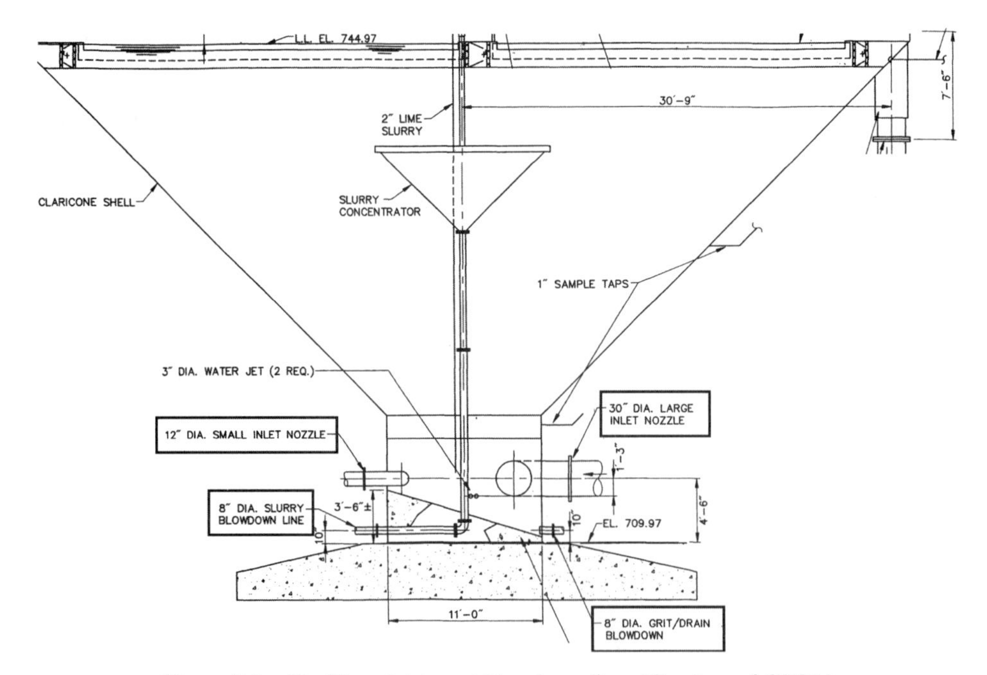

Figure 8-1 ClariCone inlets and blowdown lines (Courtesy of CB&I.)

operated in the flow range (Q) of 19,000 to 26,000 m^3/d. This yields overflow rates in the range of 2.7 to 3.7 m/h.

The 0.76 m and 0.3 m influent pipes enter the bottom cylinder tangentially to impart a spiral flow to the sludge blanket that is formed in the body of the softener (Fig. 8-1). In this blanket, precipitation, coagulation, flocculation, and particle growth must take place. Dense blankets, such as those produced by lime softening, require more tangential energy input to maintain the desired spiral flow. Since higher influent flows increase kinetic energy inputs, ClariCones can be operated at flow rates that impart high energy inputs without causing washout (overflow) of fine particles.

ENERGY INPUT TO IMPART SPIRAL FLOW

The distribution of the flow between the two inlets is adjusted using butterfly valves. This allows for control of the input of the hydraulic energy required to promote an optimal spiral flow pattern. Too little energy input causes the spiral motion to *stall* so that the suspended matter in the blanket rises vertically. Such short-circuiting results in reduced residence time in the slurry blanket and the appearance of surface upwellings or *boils* on the blanket surface.

Too much energy input can also result in the formation of a cone of depression in the center of the slurry blanket so that its periphery rises against the outer circumference. A major part of the control of the softener is to recognize when the spiral motion of the slurry blanket is near optimal.

When suboptimal conditions are observed, the butterfly valves on the influent lines can be adjusted to restore the spiral flow pattern. Throttling the butterfly valve to the larger pipe will divert more flow through the smaller pipe and increase both its influent velocity and its

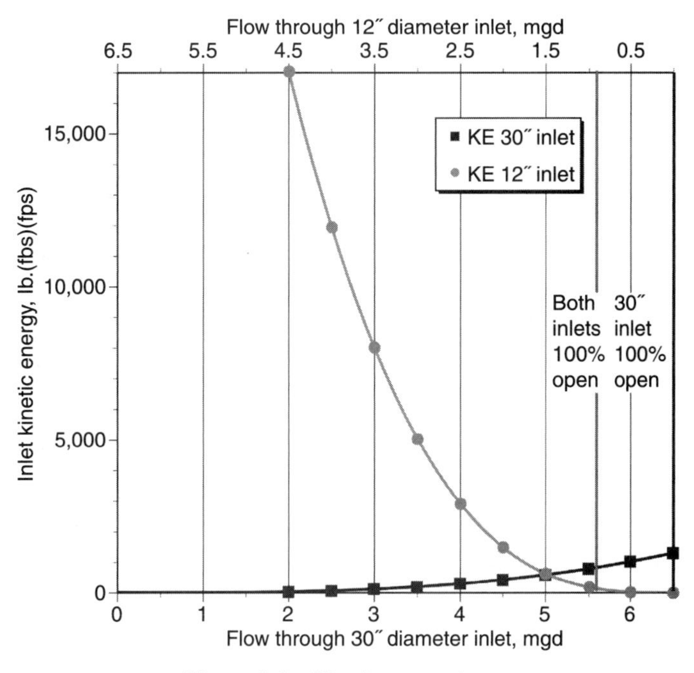

Figure 8-2 Kinetic energy input.

energy input. At low total influent flow rates, much of the flow may have to be diverted through the small inlet pipe to maintain the spiral rotation of the blanket. While butterfly valves are often operated either fully opened (100%) or fully closed (0%), the main flow valves to all operating ClariCones at Bloomington are closed to the 30% to 35% range.

In Fig. 8-2, the kinetic energy inputs through the 0.76 m and 0.3 m inlets are shown for a softener receiving a total influent flow of 25,000 m^3/d. This plot illustrates the effect of throttling the flow through the larger inlet thereby forcing an increased flow through the smaller inlet. If the larger butterfly valve were 100% open and the smaller valve 0% open (i.e., closed), the kinetic energy input would be 53 kg $(m/s)^2$. If both valves were fully open, the energy input would decrease to about 41 kg $(m/s)^2$. However, if the larger valve were throttled to reduce its proportionate flow to 15,000 m^3/d, the energy input would increase to 131 kg $(m/s)^2$.

MODIFICATION OF KINETIC ENERGY INPUT

Kinetic energy (KE) = $0.5\,mv^2$, where mass, $m = rvA$; r = fluid density = 1 kg/liter. For a pipe with a cross-sectional area, A, carrying water, KE = $0.5rAv^2$. For a 0.76 m pipe, $A_{30} = 0.44\ m^2$; for 0.3 m diameter, $A_{12} = 0.07\ m^2$; ratio $A_{0.76} : A_{0.3} = 6.25$.

In normal operation at Bloomington, the 0.3 m valve always remains 100% open while the 0.76 m valve is throttled to 60% to increase flow through the smaller inlet. For the present evaluation of the effect of kinetic energy input on slurry blanket rotation, the larger valve was further restricted by closing it, first to 35%, then to 30% of full opening. The results of these adjustments were monitored using a velocity flow meter (Fig. 8-3) positioned (Fig. 8-4) in the slurry blanket. With increased throttling of the flow through the larger inlet, the velocity was found to increase leading to visible rotation striations and an apparently stable blanket surface (Fig. 8-5).

Figure 8-3 Velocity flow meter.

Figure 8-4 Vertical positioning of flow meter.

Figure 8-5 Smooth striated surface indicates spiral flow.

EFFECT OF GRITTING ON SLURRY BLANKET

Both the periodic removal of grit and the blowdown of the solids that have accumulated in the cone-shaped slurry concentrator (positioned just beneath the slurry blanket surface) have the potential for interfering with the hydraulics of the slurry blanket.

Previous studies of the operation of the softener had shown that most of the accumulated heavy grit (Fig. 8-6) in the bottom cylinder is removed during the first two minutes of operation of the $4\,m^3/min$ grit pump. Prolonged gritting times at the high pumping rates required for effective grit removal tend to also remove significant portions of the overlying, lighter slurry blanket, exposing the concentrator cone and diluting the recovered grit.

Bloomington's most recent experience has shown that shorter and more frequent periods of gritting cause less disruption of the blanket.

Figure 8-6 Dense (dark) grit in bottom of Imhoff cones.

EFFECT OF BLOWDOWN ON SLURRY BLANKET

Similarly, the removal of solids from the concentrator cone has the potential for adversely affecting the slurry blanket and cone hydraulics. During normal operation the concentrator cone should be submerged and not visible. It should be located below the surface of the slurry blanket.

Solids from the blanket should have accumulated and compacted to a higher density with depth within the concentrator cone. Primarily, only the contents of the concentrator cone (5 m^3) should be evacuated as the overlying blanket would be expected to be less dense than the accumulated solids. To maintain a constant slurry blanket level in the softener, the frequency of the removal of the concentrator cone contents should be directly related to the rate at which solids are being formed in the slurry blanket.

FREQUENCY OF INTENSE CLEANING

After a period of time, depending upon the influent water quality and individual clarifier loading, Bloomington maintenance personnel dewater each softener for intensive cleaning. Considerable lime solids accumulations require both pressure washing of the cone sides and the physical removal of densely compacted solids from the bottom cylinder, often by bucket. Failure to routinely conduct this cleaning may result in solids accumulations that cause the upward deflection of the inflow, leading to short-circuiting and surface boils. Figures 8-7 and 8-8 illustrate the cleaning using high pressure hoses.

Figure 8-7 Pressure washing.

Figure 8-8 Washing of solids from grit settling compartment.

MAINTENANCE OF LIME DELIVERY SYSTEM

One of the most difficult and unpredictable aspects of lime softening operation relates to the lime slurry feed system. Lime slurry is pumped through a recirculating loop of interconnected flexible hose and PVC pipes. After entering the softener building, the lime feed loop is suspended from the ceiling and passes over each of the four softening units. A PVC-valved tee connector attaches a flexible hose to an automatically controlled pinch valve that throttles the flow of lime slurry to each softener (Fig. 8-9).

Figure 8-9 Lime slurry pipe loop.

Figure 8-10 Partial blockage of lime feed pipe.

Accumulations of lime (and grit) can cause blockages anywhere in the lime feed system (Fig. 8-10). Operators detect blockages when they observe increased lime feed pump pressures. Maintenance personnel are then called to determine the location of and to relieve the blockage. Since rapid restoration of lime feed often requires the replacement of pipe, spare sections must always be available.

There must be an immediate response to blockages in the lime feed line if the lime slurry blanket is not to be lost from the softeners and the solids passed onto the filters. Recent experience has shown that the slurry blanket can be essentially lost in a period as short as one and one-half hours.

Once the slurry blanket has been lost, not only will turbidities increase in both the filter influent and effluents, but the possibility of finished water turbidity exceedances will remain for up to eight hours until a stable, new slurry blanket is formed.

That the loss of the slurry blanket has been so infrequent in recent years is a tribute to the ability of Bloomington's maintenance staff to find and correct blockages before process upsets occur.

9

GRANULAR ACTIVATED CARBON

BENEFITS OF GRANULAR ACTIVATED CARBON

In the late 1980s, the City of Bloomington experienced several taste-and-odor episodes. These episodes were associated with drought, excessive algal growth, and the appearance of zones of anaerobic conditions in the lakes. Contributing factors included recurring lime feed failures, insufficient treatment unit capacities, and the lack of capabilities for removing organic compounds. When powdered activated carbon (PAC) and potassium permanganate failed to control tastes and odors, the water utility decided to install granular activated carbon caps on the filters. While GAC caps were added to all filters in 1994, primarily for taste-and-odor control, ancillary benefits included the reduction of pesticides, herbicides, and unidentified organic compounds.

In the absence of consumer complaints since the late 1980s, the city's need for taste-and-odor control was thought to have lessened. The installation of destratifiers in both lakes had provided mixing and improved source water quality, additional softening units had been constructed, and new lime slakers had improved lime feed performance.

Taste and Odor in Source Waters

In particular, the destratifiers in both Lake Bloomington and Evergreen Lake contributed to major improvements in influent water quality. Algal populations were diminished, and there were fewer of the blue-green algae commonly associated with tastes and odors. However, destratifiers remain subject to power loss and motor breakdowns that result in temporary taste-and-odor episodes.

Geosmin and 2-methylisoborneol (MIB) are often cited as the primary taste-and-odor-causing compounds in surface waters. Since analyses for MIB and geosmin are expensive

Water Treatment Plant Performance Evaluations and Operations. By John T. O'Connor, Tom O'Connor, and Rick Twait

($350 per sample as of 2007), they are not routinely monitored in most lake or treated waters. Accordingly, there is a scarce database of these measures on which to evaluate treatment process effectiveness. Instead, the threshold odor number (TON) test is one of the principal methods used to quantify taste and odor in drinking water. The TON is defined as "the greatest dilution of sample with odor-free water yielding a definitely perceptible odor" (APHA, 2006). Highly subjective, this TON test is usually conducted by a panel of tasters.

GAC Performance Warranty

The warranty provided by Calgon Carbon in their Potable Water Service Agreement stipulates that the Filtrasorb GAC "will perform the function of adsorption of dissolved organics which contribute to taste and odors during the life of the warranty . . . the threshold odor number . . . shall not persistently exceed two (2) in the carbon-filtered plant effluent for five consecutive days with the treatment plant in normal operation at the time during the first 48 months."

Total Organic Carbon Reduction

In order to comply with the requirements of the Surface Water Treatment Rule, the Bloomington water plant must remove 25% of the lake water TOC. In 2003, the plant removed an average of 56%, far beyond the SWTR requirement. The majority of TOC in the lake water was in particulate form and was removed by the physical particle removal processes rather than adsorption on the GAC.

Since GAC adsorption capacity is exhausted within months, it is assumed that longer-term removal of dissolved organic matter occurs as the result of biodegradation by the attached biological community that colonizes the carbon. Based on limited testing in the summer of 2002, a GAC-capped filter containing two-year-old carbon removed an average of 10% of the filter influent dissolved organic carbon, presumably by biodegradation, while a filter containing fresh, new carbon removed 30% of the filter influent TOC by adsorption.

Pesticide and Herbicide Reduction

Data on atrazine and acetochlor from January 1998 to June 2000 showed an average of 1.45 μg/l of atrazine and 0.29 μg/l of acetochlor in the plant influent water. While both average concentrations are well below established limits, atrazine peak concentrations in the raw water approached the maximum contaminant level (MCL) for drinking water (3 μg/l) in the summer and slightly exceeded it on one occasion. However, based on 31 sets of analyses, removals of both atrazine and acetochlor during treatment were found to average 66%.

COSTS OF CARBON

Under their 2007 contract, the City of Bloomington pays Calgon $132,300/year for the use of GAC. This includes pickup of used GAC and delivery of 73,000 kg of new GAC. Municipal staff is responsible for labor associated with the annual GAC changeout.

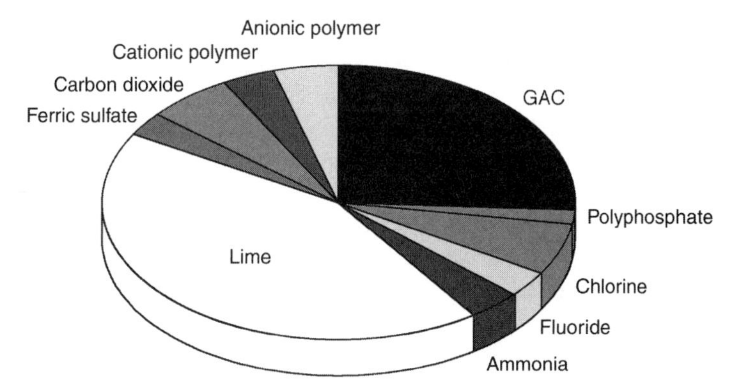

Figure 9-1 Distribution of annual chemical costs (FY '03–'04).

During a changeout, six city employees work for approximately 108 hours for a total of 648 person-hours. Assuming total labor costs of $50/hour, this represents a cost of $32,400. Since an estimated 25% of each worker's labor is spent on other concurrent projects, $24,300 is considered the changeout labor expense. Divided by seven filter changeouts per year, labor costs amount to $3,471 per filter.

As a comparison, the Bloomington Water Department received a bid of $9,573 for changeout of one filter from a local contractor (May 1999). This included labor and the use of a vacuum truck.

At $132,300, the cost of GAC accounted for 26% of the water utility's total annual chemical cost of $510,751 in (FY '03–'04) (see Fig. 9-1).

Alternatives

Maintain Status Quo There are no overwhelming reasons to change from the existing situation. The contract with the GAC supplier can be renewed annually.

Change GAC Less Frequently The GAC supplier's warranty stipulates a four-year performance guarantee for its taste-and-odor removal capability. Instead, GAC is changed out every two years in the new filter building and every three years in the old filter building. If GAC could be replaced on a four-year schedule, the annual cost of GAC would be cut nearly in half.

GAC on Fewer Filters Several filters could be designated as GAC-capped filters for use as needed. However, this might result in significant operational difficulties. Moreover, when taste-and-odor removal is urgent, it will be needed in all of the water being filtered.

Powdered Activated Carbon as Supplement PAC feed facilities are currently available at the treatment plant to assist in combating occasional episodes of taste and odor. This supplement would be more effective if fed at the Lake Evergreen intake, as there would be far longer contact time for the slow adsorption of odorous compounds to take place.

Different GAC Suppliers Alternate suppliers of GAC might be considered. This has the potential for cost savings, but carbon should not be selected based on cost alone. Various manufacturers produce many types and grades of carbon, and the quantification of each carbon's adsorptive capacity can be time consuming. Ideally, a rational basis for selection (e.g., high specific activity) should be defined, and carbon should be pilot tested under plant operating conditions over time. This would encompass both the long-term performance of the carbon and the effects of seasonal temperature changes.

Buy versus Rent The City of Bloomington is currently renting GAC from a supplier. The direct purchase and eventual resale or regeneration of their used carbon might be more cost effective and should be investigated.

REFERENCE

APHA (2006). Standard Methods for the Examination of Water and Wastewater. *American Public Health Association*, Washington, D.C.

10

PLANT OPERATIONS MANUAL

DEVELOPMENT OF OPERATIONS MANUAL AND GUIDELINES DOCUMENTS

Continually under development, the *Bloomington Water Treatment Plant Operations Manual* was created to serve as an introduction to treatment plant configuration and operation for newly-recruited plant personnel. It includes a series of illustrated, single-sheet *guideline documents* prepared to supplement the *Plant Operations Manual*. The inclusion of this specific manual is intended to serve as an example for the development of similar operator-oriented manuals at other utilities. It begins with a brief description of the history of Bloomington's water supply, its watershed, and the development and protection of its water resources.

BLOOMINGTON WATER SUPPLY HISTORY

Lake Water Sources

In 1929, owing to the growth of Bloomington to a population of 30,000, Money Creek was impounded, creating Lake Bloomington. This drinking water source replaced the groundwater supply that had served the city since 1875. A treatment plant was constructed at Lake Bloomington to produce 15,000 m^3/d (4 mgd) of potable water for the City of Bloomington. In 1971, Evergreen Lake was created to supplement Bloomington's water supply (Fig. 10-1).

Water Treatment Plant Performance Evaluations and Operations. By John T. O'Connor, Tom O'Connor, and Rick Twait

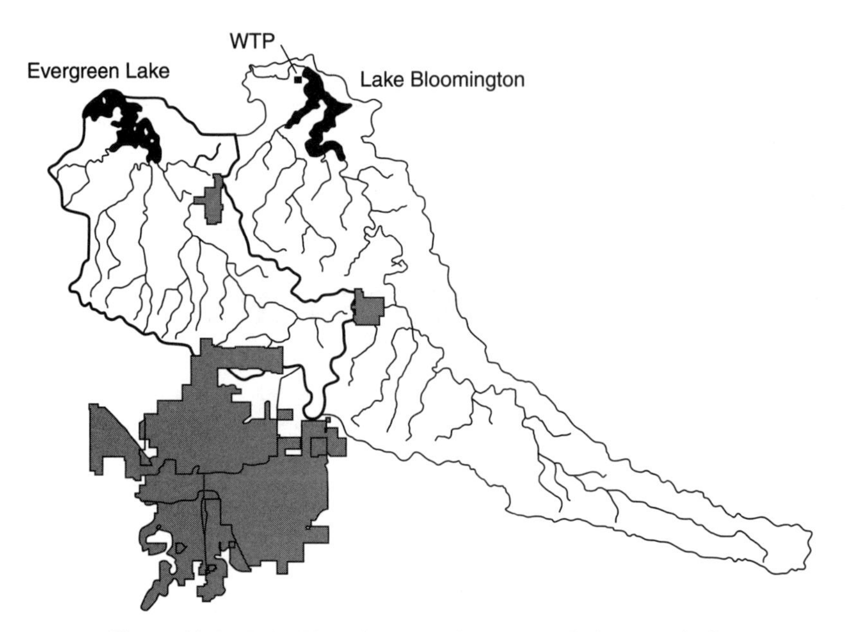

Figure 10-1 Lake Bloomington and Evergreen Lake watersheds.

Lake Bloomington and Plant	Evergreen Lake
Constructed in 1929	Constructed in 1971
Lake raised 1.5 m in 1954	Lake raised 1.5 m in 1995
Surface area: 2.6 km^2	Surface area: 3.6 km^2
Storage: 11×10^6 m^3	Storage: 19×10^6 m^3
Drainage area: 181 km^2	Drainage area: 107 km^2
Drainage to surface area ratio: 70 : 1	Drainage to surface area ratio: 30 : 1
Low-lift pumps (3): 0.5, 0.6, 0.66 m^3/s	Low-lift pumps (3): 0.4, 0.3, 0.6 m^3/s
Chemical feed: cationic polymer, PAC	Chemical feed: cationic polymer, PAC

Figure 10-2 Application of nitrogen to soil.

Figure 10-3 Animal wastes in runoff.

In 2007, to serve a daily average of 42,000 m^3/d (11 mgd) to a population of 70,000 (with 25,000 service connections), the Bloomington Water Treatment Plant (rated capacity, 84,000 m^3/d) derives its influent from two water supply reservoirs, Lake Bloomington and Evergreen Lake. Under drought conditions, a 76,000 m^3/d off-channel pumping pool on the Mackinaw River can, under stringent regulatory permit conditions, also help to replenish water in Evergreen Lake.

To control plant influent water quality, particularly with respect to *nitrate ion*, which must be held within 10 g N/m^3 (10 mg N/l) in the plant effluent, influents to the Bloomington Water Treatment Plant are a changing blend of the two lake waters.

As shown in Figs. 10-2 to 10-4, both Lake Bloomington and Evergreen Lake receive drainage and animal wastes containing nutrients (nitrogen, phosphorus) that support a wide diversity of microbial growth. Inorganic (silt, clay) and organic particulate matter (fiber, plant detritus) may range from abundant to minimal in the lakes, depending on

Figure 10-4 Algal growth in ponds.

Figure 10-5 Snow and ice cover on Lake Bloomington.

antecedent precipitation and surface runoff (lake inflow), lake level, wind, degree of mixing (lake turnover), temperature stratification, and, possibly, boat traffic. Lake water temperatures vary widely throughout the year approaching 0°C under winter conditions (Fig. 10-5).

Lakes in central Illinois periodically experience extensive growths of algae (*algal blooms*) caused by the influx of high levels of nutrients. Nutrients can enter the lake in stream inflow or be released from the lake bottom sediments (*benthos*) during quiescent periods when no oxygen is present in the lower depths.

Taste-and-odor-causing compounds can be produced by the algae or by decomposition processes occurring in the lake in the absence of oxygen (*anoxia*). Figures 1-3a–j (see color insert) illustrate both filamentous algae and inorganic particles commonly found in the lake water sources.

Heavy rainfall and surface runoff that contribute to soil erosion in the watershed can lead to rapid changes in the turbidity of the plant influent water. The rapidly-changing influx of suspended matter may require plant laboratory jar testing to determine *optimum chemical dosages* for effective coagulation. Effective coagulation results in most of the micrometer-sized particles becoming embedded in a *settleable* floc.

Lake Destratifiers

In an effort to limit the release of nutrients from the benthic sediments and to prevent the formation of anoxic zones in the lakes, innovative *destratifiers* were installed in both lakes in 1996 (Figs. 10-6 to 10-8). Lake water monitoring and laboratory analyses have shown that these units have been effective in preventing anoxic zones in both lakes. The destratifiers have also helped prevent the taste-and-odor episodes that commonly follow lake overturn in the fall.

The design, installation, and effectiveness of Bloomington's Venturi aspiration aeration systems have been documented by the Illinois State Water Survey (Raman et al., 1998).

Each system consists of a two-stage submersible pump with a rated capacity of 4.5 m^3/min. at 0.3 MPa driven by a 30 kW, three-phase 460 v, 60 cycle electric motor. This pump plus the Venturi nozzle is mounted on a skid placed on the lake bottom near

Figure 10-6 Local fisherman near destratifier at Lake Bloomington.

each raw water intake structure. The Bloomington system is small, quiet, and requires no housing or accommodations for heat dissipation. In 1996, the system for each lake cost $55,000, installed.

While lake aerators and destratifiers are widely used throughout Illinois, scientific evaluations of Bloomington's unique destratification system have shown it to be durable, versatile, aesthetically nonintrusive, exceptionally economic, and highly effective for algae control.

Watershed Protection

The City of Bloomington participates in the *Illinois EPA Volunteer Lake Monitoring Program* and has prepared *Phase I Illinois Clean Lakes Diagnostic/Feasibility Studies*

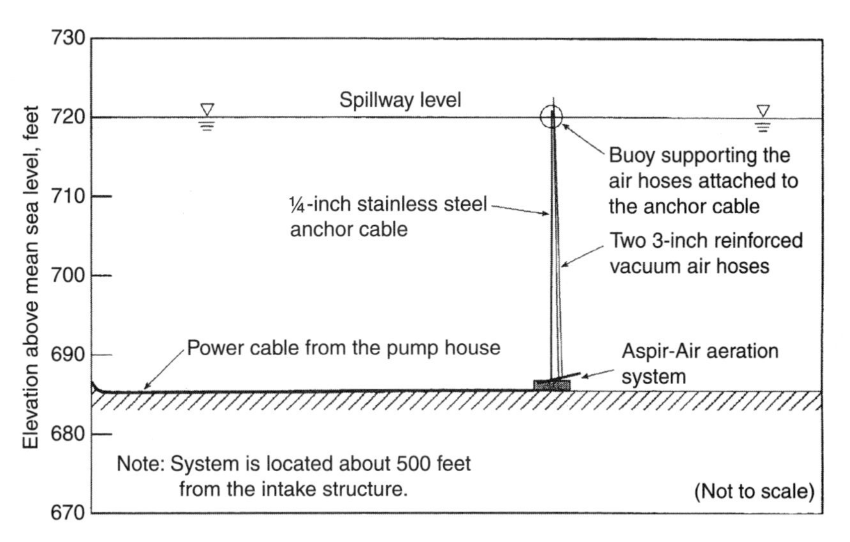

Figure 10-7 Aeration system in Evergreen Lake.

(a) (b)

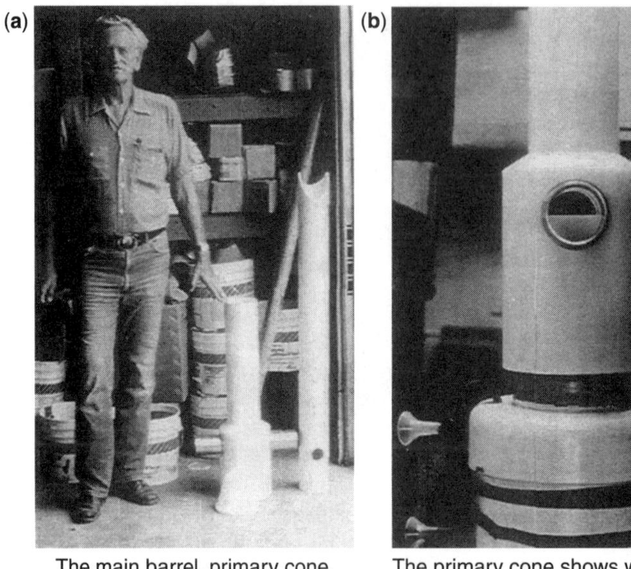

(a) The main barrel, primary cone, and directional nozzle

(b) The primary cone shows within the air chamber of the assembled unit

(c) (d)

(c) The completely assembled system, except air hoses, prior to installation

(d) System installation in the lake using a helicopter

Figure 10-8 Destratifier installation.

for both lakes. Figures 10-9 and 10-10 show the pontoon boat and sampling equipment used in the lake monitoring program. Locations for shoreline erosion control in both reservoirs were identified and stabilization structures were designed. Similarly, tributary streams draining into the reservoirs were inventoried for stability. Problem areas and stabilization measures were identified.

In addition to the in-lake activities, the city has devoted great attention and effort to limit the nutrients, particularly nitrates, originating in the drainage basins from entering the lakes. A long-term, tile drainage and nitrogen application research program has been undertaken

Figure 10-9 Pontoon boat for in-lake operations.

with Illinois State University. An experimental wetlands study with the University of Illinois is in its first phase. The city's watershed conservationist coordinates the local sponsor activities for a privately funded (Sand County Foundation) nutrient management program for the Lake Bloomington watershed. Figure 10-11 shows some of the apparatus used for measuring stream flow and depth.

Overall, the protection of the lake watersheds, including erosion control and the reduction of nutrient influxes, are vital steps in maintaining the influent quality of Bloomington's drinking source water. A range of strategies, such as lake destratification for control of algal growths plus the operational blending of lake waters to limit nitrate concentrations, are used to ensure the best possible water plant influent quality. Long-term efforts to encourage soil conservation plus research and educational programs to reduce fertilizer loss in agricultural runoff are also underway. Source water quality protection is viewed as a major part of Bloomington's water system operation.

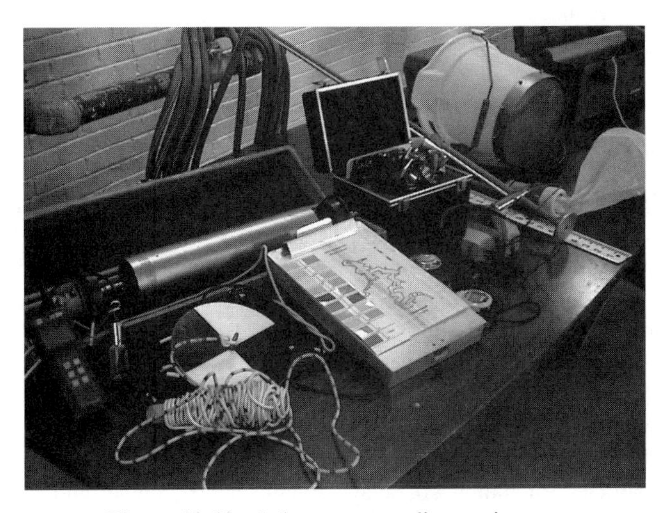

Figure 10-10 Lake water sampling equipment.

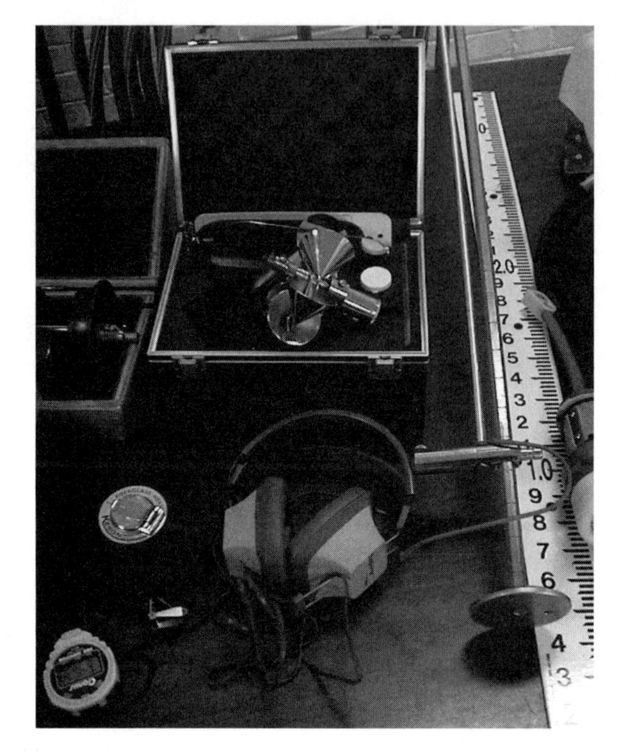

Figure 10-11 Stream flow and depth measurement apparatus.

WATER TREATMENT PLANT PROCESSES

To accommodate increasing potable water demands, the City of Bloomington has periodically increased the hydraulic treatment capacity and upgraded the sophistication of its treatment process. By 1985, the filtration capacity of its original 15,000 m^3/d plant (Fig. 10-12) had tripled in several stages to 45,000 m^3/d. Construction of the new Main Process Building (Fig. 10-13) in 1987 with six modern dual-media filters brought total plant capacity to 83,000 m^3/d.

Blend of Influent Lake Waters

The choice of plant influent water withdrawal is governed largely by lake water quality and pumping costs. While Lake Bloomington periodically exhibits high nitrate concentrations that benefit from dilution by low nitrate Evergreen Lake water, pumping from Lake Bloomington is less costly because of its proximity to the treatment works. Low lake levels are to be avoided, however, to maintain the recreational value of Lake Bloomington. In some instances, low lift pump or valve failure will limit the choice of plant influent water. Normally, pressure at the inline mixer is maintained in the range of 50 to 70 kPa, which is an indicator of energy being put into the system.

Once withdrawn, the lake waters receive comprehensive treatment for the removal of source water particles, including silt, clay, organic debris, algae, and bacteria. Beyond that, the Bloomington plant achieves significant reductions in hardness and organic

Figure 10-12 1929 Annex (old) Building.

matter. Finally, before distribution, the softened, settled, stabilized, and filtered water is disinfected to inactivate any virus particles and pathogenic organisms that have not been removed by physical treatment.

Treatment Processes

The physical and chemical treatment processes may be divided into the following ten unit operations.

Polymer Addition Upon withdrawal at their sources, the lake waters are both treated with a low dosage ($\sim 1\ g/m^3$) of *cationic polyelectrolyte*. At the Evergreen Lake water intake,

Figure 10-13 1987 Main Process (new) Building.

this polymer (coagulant) addition can be dispersed and mixed during transmission, resulting in the initial flocculation and aggregation of particles before the water reaches the inline static mixer at the treatment plant.

Coagulation Ferric sulfate is the principal inorganic coagulant used in treatment. It is applied to the influent water at the static mixer. An iron salt is used because it remains insoluble at the high pH used in lime softening. An additional effective coagulant is formed in the lime softening basins owing to the precipitation of magnesium hydroxide.

Lime Softening Lime slurry is introduced into the bottom of each softening unit where it blends with the ferric sulfate-coagulated water. Softening occurs by precipitation of calcium carbonate and magnesium hydroxide as the mixture spirals upward in the inverted cone-shaped tank.

Flocculation *Flocculation* (collision and aggregation of particles) occurs primarily in the softener's *slurry blanket.* Contact with larger, previously-precipitated particles accelerates the aggregation of the incipient precipitates.

Sedimentation Sedimentation occurs throughout the slurry blanket as particles grow in size and density. Clarification is most complete in the uppermost section (clarification zone) of the ClariCone where overflow (rise) rates are lowest owing to the increasing cross-section of the inverted cone.

Recarbonation Stabilization is accomplished in the recarbonation basins though the initial addition of *carbon dioxide* to lower pH and the subsequent application of *sodium hexametaphosphate* to sequester calcium ion and terminate the continued precipitation of calcium carbonate. Along with further settling in the recarbonation basin, this treatment reduces the solids content (turbidity) of the water applied to the filters.

Filtration While *filtration* is often thought of as the primary particle removal process in a treatment plant, only a small fraction of the particles entering or produced during water treatment are actually removed by the filters. By far, most of the particles originating in Bloomington's lake waters are removed in the coagulation, lime softening, and clarification processes. Since the filters act largely as effluent *polishers*, the amount of solids applied to them should be minimized to allow for the longest and most effective filter runs.

However, evolving federal and state regulations have caused utilities to focus more attention on the performance of individual filters by installing continuous flow *turbidimeters*. At Bloomington, system control and data acquisition (SCADA) systems continuously display and record effluent turbidities for each individual filter.

The filters at both plants are *capped* with 1 mm granular activated carbon (GAC) over 0.5 mm filter sand. The GAC is in place primarily for *adsorption* of taste-and-odor-causing *solutes*, while the sand is for particle removal and protection against the passage of microbial pathogens. However, the solutes adsorbed on the GAC may be converted to microbial cell mass, thereby leading to the development of a layer of bacteria on the media surface.

Accumulated particles and microbial accumulations are removed from the filters through *backwashing* with *finished* water. The *backwash water* can be sent directly to the sludge settling lagoons or it can be recovered to the *head* of the plant in a storage and settling tank called the *reclaim basin*. From the reclaim basin, the settled backwash water (*supernatant*) is returned (recycled) to the plant inflow at the inline static mixer.

After chlorine and ammonia are added for *secondary disinfection*, the filter effluents are discharged to *clearwells* and then flow through conduits into the 7,600 m^3 *finished water storage reservoir*. The finished water storage reservoir must provide sufficient retention time for the disinfection of the water to be complete before final distribution of the treated water.

Disinfection *Primary disinfection* is accomplished at the Bloomington water plant by the addition of chlorine to form a *free* (hypochlorous acid) *chlorine* residual. State and federal regulations require a disinfectant concentration multiplied by time ($C \times t$) to achieve a defined inactivation of pathogenic organisms. Bloomington's overall treatment routinely results in far more than the required reduction of organisms of human health concern.

Chloramination *Ammonia* and *chlorine* are both added following filtration to yield a secondary, *chloramine*, residual. This less reactive form of disinfectant makes it possible for Bloomington to maintain a persistent residual throughout its entire distribution system for residence periods of a week or longer. Chloramine is particularly effective in suppressing *microbial regrowth* in water distribution systems, thereby avoiding regulatory violations due to *coliform* and *heterotrophic plate count (HPC)* organisms. Chloramine is the disinfectant residual that arrives at the consumer's tap.

Fluoridation Fluoride is added to Bloomington's finished water to bring its total concentration up to 1 g/m^3.

Water Treatment Flow Diagram

Figure 10-14 outlines the hydraulic balance and treatment processes employed at Bloomington. Source water is first pumped from Lake Bloomington and/or Evergreen Lake to a 0.76 m inline static mixer where the primary coagulant, ferric sulfate, is injected. Up to 2,600 m^3/d of reclaimed backwash and/or sludge lagoon supernatant may also be blended with the influent at this point.

The chemically-treated, blended water is then conveyed through a *meter vault* to a preengineered *main clarifier building* that houses four upflow solids-contact softener/clarifiers (ClariCones). Following blending of lime slurry with the ferric sulfate-coagulated plant influent in the base of the ClariCones, particle removal, softening and settling occur in these large, inverted cone-shaped basins which feature slurry blankets to aid in the complete entrainment of particles in the source water. This state-of-the-art facility was placed in service in 2000.

The lime-softened and settled water from the softening units then flows into two, 18 m diameter *recarbonation basins* with serpentine effluent launders.

The CO_2-stabilized water is then directed to the 12 GAC-capped sand filters at the 1929 Annex Building and/or the six GAC-capped sand filters at the Main Building. To improve

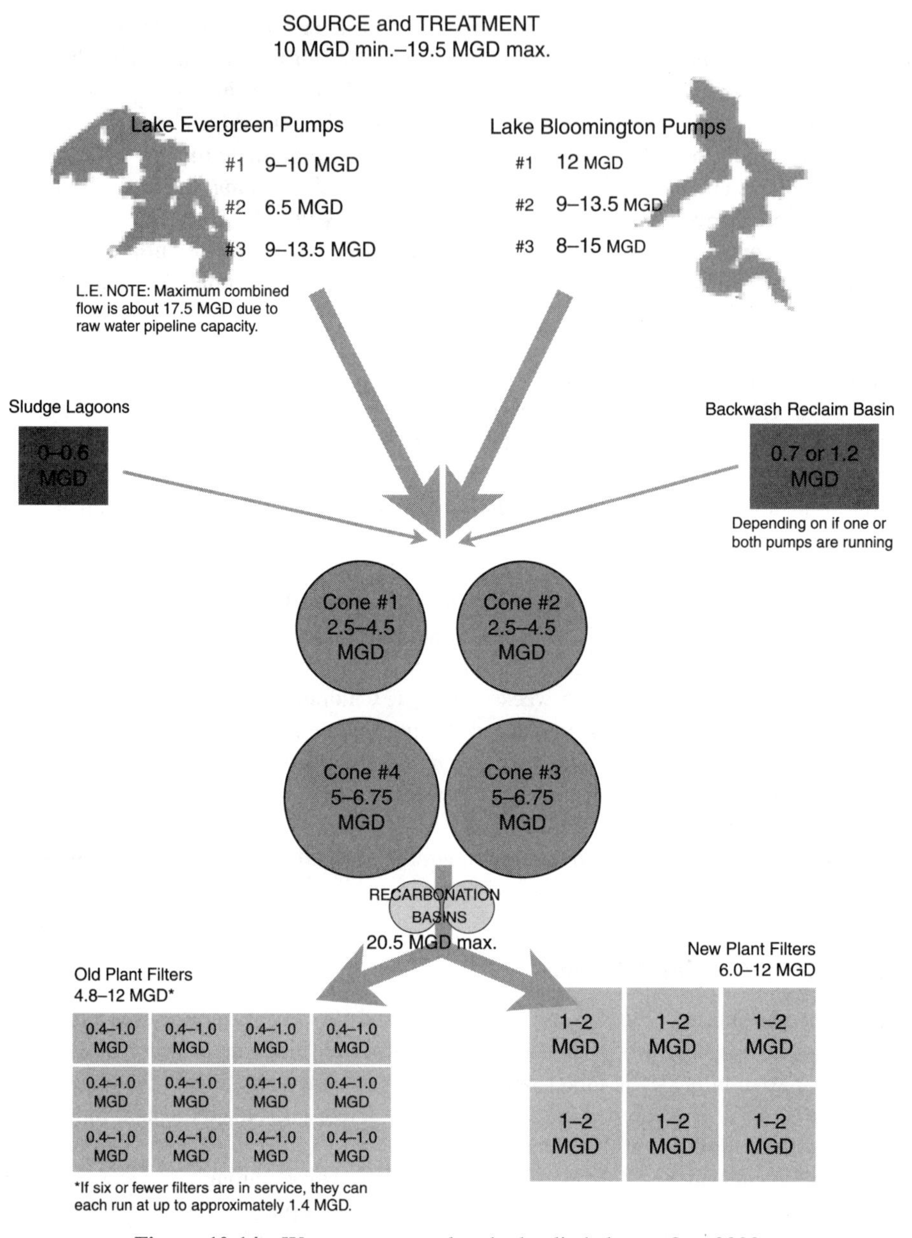

Figure 10-14 Water treatment plant hydraulic balance, Oct. 2008.

plant efficiency, softened backwash water from each of the filters is transferred to a 1,500 m^3 reclaim basin. From there, it is reintroduced into the plant influent flow at a slow rate.

Aqueous solutions of chlorine and ammonia gas are introduced prior to storage of finished water in a 7,600 m^3 underground finished water storage reservoir. Together, these chemicals form chloramine, the final disinfect residual that will persist in the distribution system. The finished water storage reservoir (Fig. 10-15) provides additional retention time for disinfection to take place before distribution.

Figure 10-15 Underground finished water storage reservoir.

Lime Storage and Slaking

Calcium oxide, CaO, also called quicklime, is delivered in bulk to the lime storage building (Fig. 10-16). It is pneumatically ejected from hopper-bottomed tractor-trailers (Fig. 10-17) into large storage hoppers as dry powder (Fig. 10-18).

Prior to lime softening, dry quicklime, CaO, must be converted to calcium hydroxide, $Ca(OH)_2$, slurry by adding water. This process, called *slaking*, generates heat and results in a pasty mix of hydrated lime and water that is difficult to meter uniformly into the water to be treated.

$$\text{Slaking reaction: } CaO + H_2O = Ca(OH)_2 + \text{Heat}$$

Approximately 120 g/m^3 of dry quicklime are applied for softening. For an average flow of 45,000 m^3/d, 5,400 kg must be slaked daily. Uniformity of the calcium hydroxide slurry formed by slaking is essential for uniform softening. Either excessively strong or weak

Figure 10-16 Lime storage building.

Figure 10-17 Delivery of lime to storage.

slurries or the loss of slurry feed results in plant upsets, affecting both the softening reactions and particle removals.

Paste-Style Slakers Prior to 2001, lime was produced using *paste-style continuous flow slakers* (Figs. 10-19 to 10-21). The lime that these paste slakers produced was conveyed immediately to the softening units. Therefore, any variation in lime quality (or loss of feed) would quickly result in lower treated water quality. Because of caking, the lime slakers, augers, and lime conduits required the highest level of daily attention, maintenance, and housekeeping.

Batch Slakers An advanced *batch* slaking system was introduced in 2001 (Fig. 10-22). The *Tekkem slaking system* addressed many of the problems experienced with the

Figure 10-18 Lime storage hoppers.

Figure 10-19 Paste-style lime slakers.

paste-style slakers. The use of load cells to control (weigh) the amount of lime and water added per batch minimizes variation in quicklime density. Temperature probes and microprocessors allow the system to automatically monitor temperature increases and be adjusted for variations in quicklime reactivity (Figs. 10-23 and 10-24).

However, the batch mode requires the maintenance of some reserves of stored lime slurry. *Slurry storage* (Fig. 10-25) and continuous recirculation (Fig. 10-26) facilitate blending of batches to achieve equalization of lime quality. It also provides Bloomington plant maintenance personnel with additional time to fix any system breakdown before treated water quality is impaired.

The new slaking system also requires a sophisticated recirculation system to deliver the appropriate flow of lime slurry to each softener and return the unused portion to slurry storage. Magnetic flowmeters monitor treated water flowing into each softener and automatically adjust the lime slurry feed applied to each unit with a pinch valve. While new operating procedures have been developed to minimize lime blockages and feed interruptions, operators report that plant upsets due to lime system problems were significantly reduced after this new slaking system was installed.

Recirculating Lime Slurry Pipeline

Lime slurry is continuously recirculated to the softener building in a loop consisting of a combination of flexible hoses and PVC pipe (Fig. 10-27). Figure 10-28 shows how

Figure 10-20 Auguer in lime slurry.

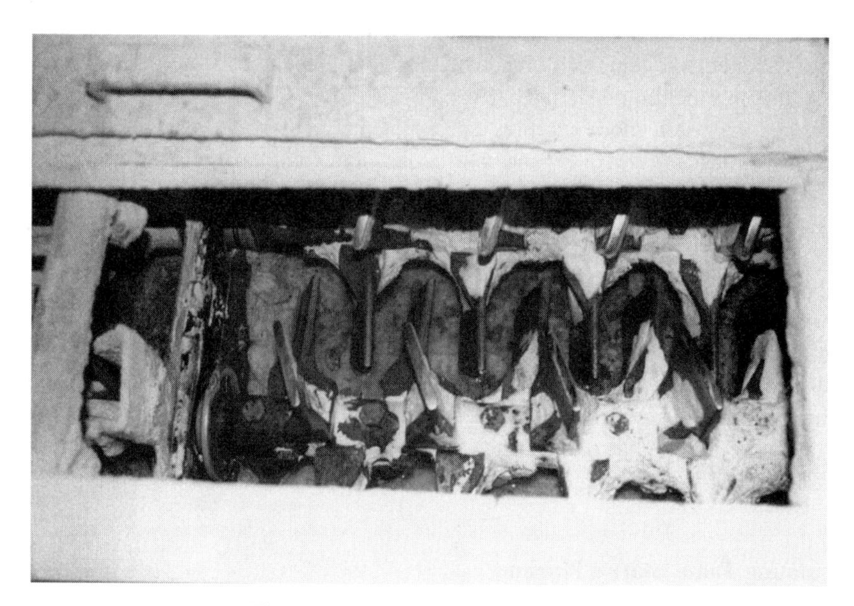

Figure 10-21 Mixing of lime slurry.

Figure 10-22 Batch (Tekkem) slaker.

the interior of the PVC pipe is gradually eroded by the hard *grit* that accompanies the lime slurry. Because of recent modifications, each individual softener now receives its lime slurry feed through a vertical hose dropped into the top of the softener (Fig. 10-29). This modification allows for the manual adjustment of the depth at which the lime slurry is discharged (Fig. 10-30) and results in greater efficiency in lime utilization.

Figure 10-23 Control panel for slakers.

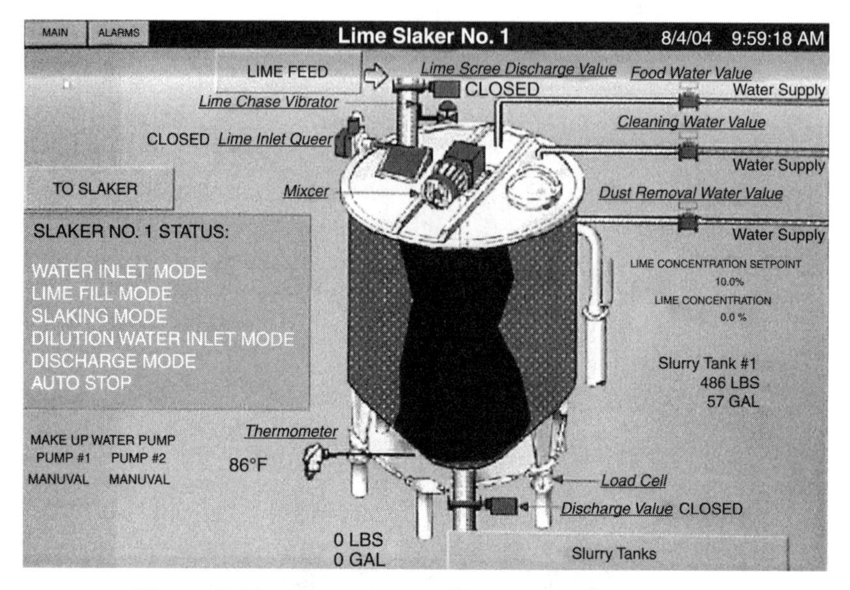

Figure 10-24 SCADA system representation of lime slaker.

Upflow Solids-Contact Softener/Clarifiers

As illustrated in Chapter 8, the proprietary ClariCone, manufactured by CBI Walker, is an upflow solids-contact softener/clarifier which allows the mixing of lime, softening, coagulation, and settling to occur within an inverted cone-shaped steel tank (see Fig. 8-1). Influent water and lime slurry are introduced tangentially at the bottom cylinder of the tank. As this mix swirls and rises in the conical section of the tank, its upward flow velocity decreases progressively owing to the increased cross-sectional area of the inverted cone.

As practiced in Bloomington, the lime softening reaction results in the precipitation of both calcium carbonate and magnesium hydroxide. The magnesium hydroxide serves as

Figure 10-25 Slaked lime slurry storage, mixing, and recirculating pumps.

Figure 10-26 Lime slurry distribution system: pumps, valves, flexible hose.

Figure 10-27 Lime slurry piping to softeners.

Figure 10-28 Grit-abraded invert of PVC pipe.

a coagulant that entrains both the particles in the influent water and calcium carbonate. Approximately 9 m above the base of the tank, the upward water flow velocity decreases to 1 mm/s, the approximate settling rate of the floc formed.

At the point where the floc settling velocity equals the *rise* or *overflow rate* of the liquid in the tank, the surface of the slurry blanket forms. Above the blanket, the water is progressively clarified as its rate of rise continues to decrease in proportion to the square of the cone diameter. The clarified softener effluent is collected in a large, adjustable, single trough (see Fig. 7-8) near the 10.5 m level where the overflow velocity has further decreased to about 0.75 mm/s.

Figure 10-29 Vertical lime slurry PVC distribution pipe.

Figure 10-30 Lime slurry flow meter and peristaltic feed pump.

The solids produced in the softener are collected from two levels. A portion of the lighter, coagulated solids from the top of the blanket enters an internal, cone-shaped *slurry concentrator* (Fig. 7-7). Some compaction occurs in this smaller cone and, periodically, its contents are withdrawn through a *slurry discharge line* at the base of the tank. In addition, there is an accumulation of heavier solids (grit) with settling velocities $>25\,mm/s$ in the cylinder at the base of the tank. Since grit is often difficult to suspend and remove, *jetting*, a brief injection of clarified water, is activated prior to (or following) opening of the grit chamber valve.

Four ClariCones are housed in an enclosed building, which also provides protection and storage for bulk chemicals and mobile equipment (Fig. 10-31).

Starting Up a ClariCone Softening units are commonly operated within a narrow range of flow rates in order to maintain the hydraulic stability of their slurry blankets. Operators either bring them fully online or remove them from service to meet significant changes in system demands. The following protocols are used at Bloomington:

- Decide which softener (20,000 m^3/d or 26,000 m^3/d) to start up based on how much the flow is to be increased. Excess *plant production* over *system demand* flows is accommodated in *finished water storage* if capacity is available.
- Increase chemical feed rates based on the increased flow through the plant.
- Increase the lake influent pumpage to meet the increased flow requirements of the softeners that will be in service.

Figure 10-31 Enclosed building houses four ClariCones.

- Start up additional filters, as necessary, to limit filter flow rates.
- After the next set of operational samples, reevaluate the adequacy of the chemical feeds.

Shutting Down a ClariCone

- Decide which softener to shut down based on desired decrease in flow and whether a particular softener needs to be cleaned. Cleaning should generally be done every two to three months.
- Reduce chemical feeds based on the reduced flow through the plant, for example,

  ferric sulfate: using SCADA, change ferric feed *output* setting (target: 1.2 g/m^3)

  ammonia: change flow manually with dial above rotameter (target: 1.2 g/m^3)

  chlorine: change flow manually with dial above rotameter (target: 4.2 g/m^3)

  fluoride: change feed manually (target: 0.9 to 1.2 g/m^3 in finished water); chemical feed pumps: stroke remains constant, speed is varied (Fig. 10-32).

  carbon dioxide: change gas flow manually with valve before rotameter (target: 5 g/m^3; effluent pH $\sim$ 10)

  hexametaphosphate: no change (feed rate remains constant at 22 kg/d)

  anionic polymer: using SCADA, shut off feed pump corresponding to softener being shut down (Fig. 10-33).

  lime: should shut itself off once flow to softener is shut down.
- Visually check to confirm all chemical feeds that were changed using SCADA.
- Decrease lake influent pumpage to meet the reduced flow requirements of the softeners remaining in service.

Most of the operational changes can be executed using the water plant's automated SCADA system. The operator will access the *SCADA Softeners Screen* that illustrates the units and their current operating conditions to make the following adjustments:

- Change flow setpoint of softener being shut down to zero.
- Modify setpoints of other softeners as necessary to handle plant influent.

Figure 10-32 Hydrofluosilicic acid bulk storage tank and feed pumps.

Figure 10-33 Anionic polymer feed to ClariCone influents.

- Initiate extended grit (30 minutes) and then a blowdown (15 minutes) on softener being shut down.
- Disable grit and blowdown on softener being shut down.
- Disable lime flushing valve of softener being shut down.
- Shut softener influent valve manually to ensure zero flow.
- Shut down filters as necessary.
- Wash any filter that was shut down.
- After next set of operational samples, reevaluate chemical feeds.

Softening by Precipitation with Lime

As noted, the Bloomington Water Treatment Plant softens water by precipitating both calcium carbonate and magnesium hydroxide with the addition of lime. The bar diagrams in Figs. 3-5 and 3-6 (electroneutrality conditions) illustrate the inorganic composition of the water before and after lime softening. In addition, the differences in the lengths of the top and bottom diagrams indicate the amount of lime softening sludge precipitated.

For the year 2000, as indicated in Table 10-1, the Bloomington plant removed 43% of the influent water total hardness (as g $CaCO_3$ equivalent/m^3) and, coincidentally, 43% of the influent total organic carbon (TOC). The TOC was removed primarily as part of the coagulation and softening processes.

For comparison, the source and finished water hardness for a number of midwestern cities is given in Table 10-2.

TABLE 10-1 Removal of Magnesium, Calcium, and TOC by Lime Softening

Sampling Point		% Removal
	Magnesium, g/m^3 as $CaCO_3$ eq.	
Influent	85	—
Finished	39	54
	Calcium, g/m^3 as $CaCO_3$ eq.	
Influent	111	—
Finished	79	29
	TOC, g/m^3	
Rapid Mix—Influent	6.10	—
ClariCone (20,000 m^3/d)	3.54	42
ClariCone (26,000 m^3/d)	3.64	—
Recarbonated Influent	3.48	—
Recarbonated Middle	3.56	—
Recarbonated Effluent	3.49	—
Filter 13 Influent	3.46	—
Filter 13 Effluent	3.46	43

TABLE 10-2 Hardness Reductions in Midwestern Cities

City	Water Sources	Hardness, g $CaCO_3$ eq./m^3	
		Raw	Finished
Kansas City, MO	75% MO River; 25% alluvial wells	218	85
St. Louis, MO	66% MS River; 34% MO River	208	107
Columbia, MO	Alluvial wells in MO River flood plain	350	155
Chicago, IL	Lake Michigan; South District Plant	128	128
Highland, IL	Silver Lake	104	141
Normal, IL	14 Wells	419	108
Bloomington, IL	Lake Bloomington, Evergreen Lake	196	111

Lime Softening Sludge

The softening sludge from the precipitation of hardness is transferred, first, to the nearby *sludge vault*, thereafter, to extensive settling ponds where the sludge slowly compresses and dewaters over weeks and months (Fig. 10-34). The overlying supernatant from the sludge settling ponds is periodically returned to the reclaim basin where it mixes with backwash water recovered from filter washing. After quiescent settling in the reclaim basin, the blend of supernatant and backwash is gradually transferred to the plant inline static mixer for resource recovery.

The *Imhoff cones*, Fig. 8-6, illustrate the dewatering (compaction) of the lime softening sludge (white) over a period of minutes and hours. The heavy grit (dark) settles to the bottom of the cones within minutes. The greatest degree of sludge compaction possible (highest percent solids; least moisture content) is desirable since the dewatered softening sludge will ultimately be hauled away by truck for disposal on agricultural land.

Figure 10-34 Lime sludge storage and dewatering lagoons.

Gritting

Evaluated and illustrated in Chapter 6, *gritting* is the process of removing the settled material from the lower cylinder of the ClariCone. If an excessive amount of heavy material (unreacted lime, calcium carbonate, sand, silt, chert, and debris from influent water) is allowed to accumulate in the bottom of the tank, it can disrupt the helical flow of influent water and lime, deflecting the flow upward, disrupting the blanket, and causing short-circuiting.

Much of the grit is acid-insoluble material that comes with the lime itself. It consists of impurities, such as silica sand and chert, as well as hard-burned lime. Since approximately 5,400 kg of lime containing as much as 5% to 8% grit are applied to the Bloomington's average daily flow of 45,000 m^3, about 350 kg of grit may be introduced to the softeners each day. For a single softener producing 19,000 m^3/d, up to 150 kg of grit per day may be introduced along with the lime feed.

At 19,000 m^3/d, the upward water velocity in the bottom cylinder of a ClariCone is 25 mm/s. Therefore, in order to settle into the bottom cylinder, a particle must have a settling velocity of greater than 25 mm/s.

Grit is automatically pumped from each operating ClariCone every eight hours. A 3.8 m^3/min. grit pump operates for five minutes discharging the accumulated slurry through a 0.2 m pipe. Nominally, 19 m^3 are discharged through the 5 m^3 cylinder volume.

To assess grit removal through a grit removal cycle, samples from a tap in the grit discharge pipe downstream from the pump were collected and analyzed for total solids. After two minutes (or 7.6 m^3 removed), gritting was withdrawing water from the bottom of the slurry blanket. This indicated that prolonged gritting might disrupt the structure of the blanket.

Recarbonation: Stabilization of Filter Influent

Settled water (clarified effluent from the upflow solids-contact basins) passes into two recarbonation basins (18 m in diameter, 4.4 m deep). The diffusion of carbon dioxide gas to the bottom of these basins lowers the pH of the water, arrests further precipitation of calcium carbonate, and dissolves some of the residual of the newly-precipitated calcium carbonate. The process of *stabilization* reduces the numbers of particles (turbidity) entering the filters, extends filter runs, and minimizes the build-up of calcium carbonate on filter walls and the grains of filter media.

Operationally, when an excess of lime (lime overfeed, pH $>$ 11+) has been fed into the softeners, dissolved calcium hydroxide in the clarified effluent will react with the carbon dioxide to precipitate *additional* calcium carbonate in the recarbonation basins (Fig. 10-35). This is seen as an increase in turbidity. Bloomington operators routinely check the clarity of the water in the recarbonation basin for evidence of *lime overfeed* (see Figs. 6-15 and 6-16).

Carbon dioxide is stored as a liquid in a horizontal tank, vaporized, and fed as a gas through two gas flow control rotameters. Also in the CO_2 control building (Fig. 10-36), sodium hexametaphosphate is metered into the effluent of the recarbonation basin at a constant rate to prevent further aggregation of precipitates prior to filtration (Fig. 10-37).

Filter Operations

The design parameters, operation, effluent monitoring, and backwash protocols for Bloomington's filters are presented in Chapter 5, Filter Operations. This chapter also

Figure 10-35 Recarbonation basin, carbon dioxide storage tank, and supply tanker truck.

discusses turbidity requirements and the characteristics of new and aged GAC and sand media. Special attention is given to TOC removal and the effects of microbial growth and oxygen depletion on filter operational requirements. Finally, consideration is given to filter surveillance, measurement of bed expansion during backwash, and the removal of microbial accumulations from filter media.

Figure 10-36 Carbon dioxide control building.

Figure 10-37 Hexametaphosphate is fed to effluent of recarbonation basin.

Primary Disinfection

As discussed earlier, the final barrier to disease-bearing organisms at Bloomington is provided by chemical disinfection (chlorination). As required by federal and state regulations, Bloomington maintains a disinfectant residual concentration, *C*, for a prescribed time, *t*, such that a sufficient destruction of microbial pathogens is achieved. Together, physical removal plus disinfection is intended to achieve a minimum of 99.99% total reduction in the most resistant pathogens.

Figure 10-38 Chlorine cylinder storage and feed controls, alarms.

Figure 10-39 Entrance to chlorine feed control room.

Chlorine is brought to the plant as a liquid in pressurized ton cylinders. The cylinders are stored in a separate enclosed room with numerous safety features (Fig. 10-38). From this room, chlorine gas is evolved and distributed to a second enclosed room (Fig. 10-39) containing the *chlorinators* (Fig. 10-40), equipment that meters chlorine gas and dissolves it in water to form a concentrated chlorine solution that can be safely applied (Fig. 10-41) to the water entering the ClariCones. In the absence of ammonia, a *free chlorine residual* is formed that provides *primary disinfection* for the water during softening, settling, and recarbonation.

Figure 10-40 Chlorine controls and feeders.

Figure 10-41 Distribution of chlorine solution.

Secondary (Final) Disinfection

Following filtration, both chlorine and ammonia solutions are added (Figs. 10-42 and 10-43) to form *monochloramine*.

$$NH_3 + HOCl = NH_2Cl + H_2O$$

Figure 10-42 Ammonia gas feed outbuilding.

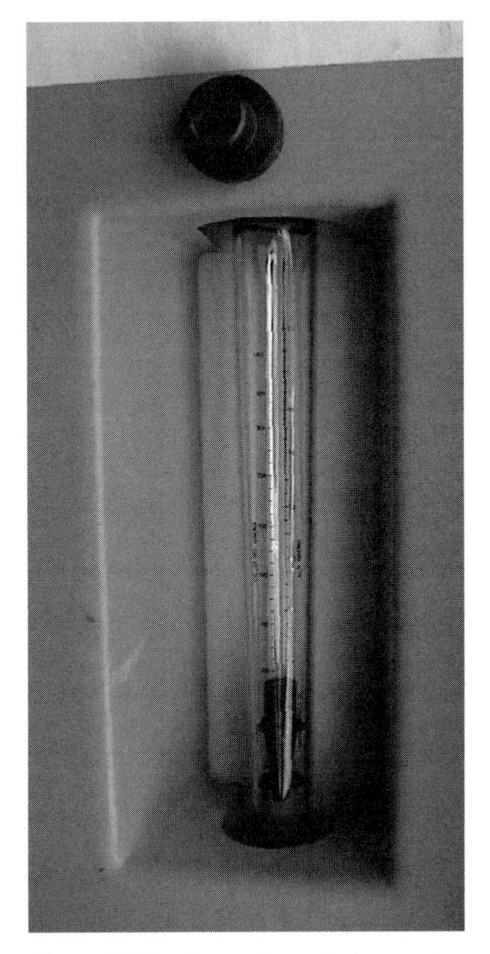

Figure 10-43 Ammonia gas feed rotameter.

Chloramines have the advantages of persistence in the distribution system and low rates of reaction with organic matter to form *disinfection by-products*. Although chloramines are slower-acting disinfectants than free chlorine, they are persistent and more effective in preventing microbial regrowth in the distribution system.

LABORATORY FACILITIES

The Bloomington Water Treatment Plant maintains excellent laboratory facilities for chemical and microbiological analyses. The prime purpose of this laboratory is to ensure regulatory compliance, such as with analyses for coliform bacteria. However, the laboratory is fully equipped to provide data to support all plant functions, assessing softening by precipitation, filter performance, disinfection, sludge solids concentrations, control of nitrate and fluoride, grit removal, and numerous additional functions as called for by plant operators.

In the laboratory, turbidity analysis can be supplemented with particle counting plus microscopic particle identification and enumeration. Taste-and-odor, a very subjective

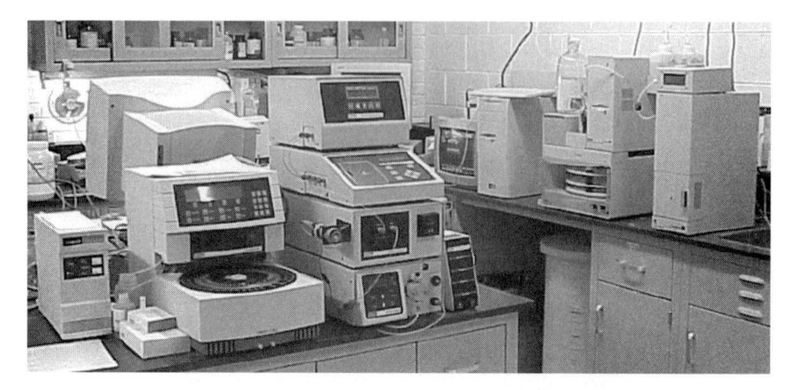

Figure 10-44 Laboratory analytical equipment.

parameter, can be supplemented with data on TOC. Hardness reduction, as well as the removal of numerous metals and inorganic species, can be accurately measured using a modern ion chromatograph.

The laboratory is regularly inspected and the analytical procedures certified by the Illinois Environmental Protection Agency. In addition to conducting the analyses necessary to demonstrate compliance with all state and federal regulations, it is a readily accessible and vital tool for Bloomington plant operators to call on in the event of plant upset or process modification (Fig. 10-44).

SHOP FACILITIES

Extensive shop and storage facilities are located in the 1929 Building (Fig. 10-45). In addition to equipment maintenance and repair, special projects are constructed in this workshop. A recent example is an advanced filtered water turbidity monitoring and recording

Figure 10-45 Bloomington Water Plant shop facilities.

Figure 10-46 Panel designed to monitor both acidified and unacidified filtered water turbidity.

facility that allows for the continuous monitoring of turbidity both with and without acidification (Fig. 10-46).

This distinction between acidified and unacidified turbidity is of particular importance for Bloomington Water since much of the plant's finished water turbidity is due to post-precipitation of calcium carbonate. Turbidity caused by post-precipitation of hardness does not constitute a health threat and, with regulatory agency (IEPA) approval, samples of finished water may be acidified to demonstrate compliance with the Surface Water Treatment Rule requirements for turbidity.

Following the completion of secondary disinfection with chloramines, the high-lift pumps (Fig. 10-47) send water to the Bloomington distribution system directly from the underground finished water storage facility.

Figure 10-47 High-lift pumps to distribution system.

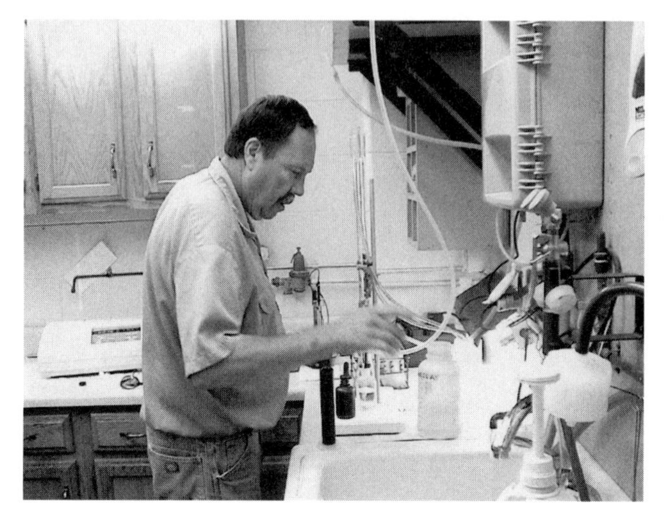

Figure 10-48 Operator's laboratory.

OPERATOR'S LABORATORY

Every two hours, the operator on duty makes the rounds to collect seven samples; one each from four softeners, two recarbonation basins, and the plant effluent (lab tap). In the Operator's Laboratory (Fig. 10-48), the following chemical analyses are performed:

- pH, turbidity, turbidity on an acidified sample—on all seven samples.
- Total chlorine residual—on plant effluent
- Alkalinity every four hours (target is >18 mg/l as $CaCO_3$ equivalent for copper/lead compliance)
- Total and calcium hardness (calculate magnesium) every six hours
- Free chlorine (hypochlorous acid) in softener effluents every six hours
- Fluoride in plant effluent
- Water temperature of plant effluent measured daily

The operator's laboratory pH meter is standardized daily.

COMMUNICATIONS BETWEEN OPERATORS AND OPERATIONAL CONTINUITY

Over the years, Bloomington operators have developed a clear, formal plus informal, system of inter-operator communication. Operators communicate recent changes in treatment, forthcoming (e.g., bulk chemical) needs, maintenance requirements, emergency response contacts and procedures, and numerous other items of operational concern (Fig. 10-49). This communication is generally documented and takes many forms, such as:

- Daily logs
- Specific task assignments (e.g., bulk chemical inventory and reordering)
- Maintenance of filtration records (new plant and old plant)

Figure 10-49 Operator change-of-shift update.

- Posting of duty rosters
- Activity checklists
- Use of a dry erase board for immediate status reports

Bloomington plant management also seeks to offer in-house operator training programs and provide regular opportunities for operators to make recommendations for operational improvement. With the treatment plant control room serving as a focal point, there are daily opportunities for operators to meet with supervisory, maintenance, and laboratory personnel. The provision of a sufficiently large and convenient area for these exchanges has been essential to effective communication among staff.

FUTURE DEVELOPMENT OF OPERATIONAL GUIDELINES

Based on the combined experience of Bloomington's senior operators, this operations manual is intended primarily to provide introductory guidelines for new water treatment plant personnel. Since the Bloomington water system is dynamic and ever-growing, its operating manual must be continually upgraded to accommodate changes based on new operational experiences, the installation of new processes and automated systems, the promulgation of new regulations governing water quality, changing economic circumstances, and numerous operational decisions and constraints that cannot be foreseen.

However, the benefit of building upon the vision, experience, and labor of those senior operators who have dedicated their working lives to the Bloomington water system cannot be overemphasized. Accordingly, this manual is directed to those who will inherit the responsibility of providing and safeguarding the city's water supply.

REFERENCE

Raman, R. K. et al. *Aeration/Destratification in Lake Evergreen*, McLean County, Illinois, March 1998.

11

TASTE-AND-ODOR CONTROL

Although infrequently since the late 1980s, Bloomington occasionally receives complaints of musty tastes and odors in its treated municipal drinking water. The purpose of this investigation was to evaluate water quality and treatment plant performance in reducing tastes and odors that evolved in the Evergreen Lake water source in 2004. In addition, current and potential future treatment alternatives for enhanced odor remediation were evaluated.

Complaints of musty odors in Bloomington's distributed water were first received in November 2004. At the time, Lake Bloomington, generally free of noticeable odor, was in use as the sole treatment plant source water. However, weeks of steady rain and snowmelt had increased both surface infiltration and runoff, resulting in the leaching of nitrate ion from agricultural tiles throughout the primarily agricultural Lake Bloomington watershed. To maintain compliance with drinking water regulations, concentrations of nitrate levels are closely monitored in Bloomington's lakes and tributary flows.

By late November 2004, both of Bloomington's lakes had filled to overflowing due to high rates of stream inflow. Lake Bloomington nitrate levels had increased to just under the *maximum contaminant level* (MCL) of $10\,g/m^3$ as nitrogen (Fig. 11-1). Projected further increases necessitated the blending of low nitrate Evergreen Lake water with Lake Bloomington water to dilute and diminish the finished water nitrate ion concentration.

Accordingly, on December 2, the influent to the water treatment plant was adjusted to a blend of 60% Lake Bloomington and 40% Evergreen Lake water. Figure 11-2 illustrates the frigid winter conditions at Evergreen Lake. While the blend effectively reduced nitrate levels, noticeable odors from Evergreen Lake water were detected in the plant's finished water within a day.

Water Treatment Plant Performance Evaluations and Operations. By John T. O'Connor, Tom O'Connor, and Rick Twait

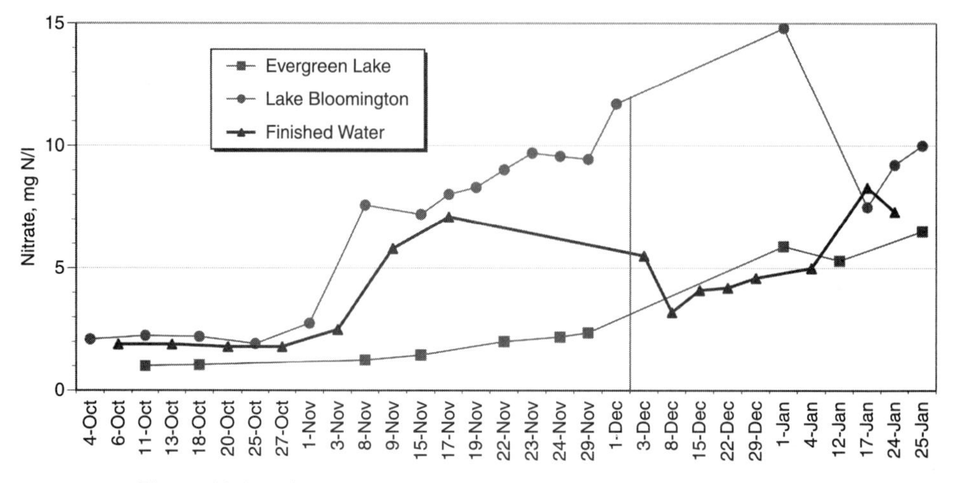

Figure 11-1 Nitrate concentration increases in lake and finished waters.

Figure 11-2 High lake stage due to rain and snow.

When the Evergreen Lake water exhibits odor, it is typically described as *musty* or *earthy*. However, this taste and odor was undetectable to most of the plant personnel who sniffed or tasted the water. This limited the pool of potential evaluators of the effectiveness of treatment techniques for reducing tastes and odors. As tastes and odors are notoriously difficult to quantify, this marked difference in olfactory sensitivity creates a continuing problem in rapidly assessing remedial treatment effectiveness.

NUTRIENT SOURCES AND ORGANISM GROWTH

Nutrients in agricultural runoff and drainage, particularly nitrogen and phosphorous, contribute to the prolific growth (blooms) of microorganisms, such as algae, fungi, and bacteria

(Figs. 1.3f to 1.3i; see color insert). These blooms are believed to be the root causes of Evergreen Lake's occasional musty taste-and-odor character. Microorganisms, such as cyanobacteria and actinomycetes, produce a wide variety of volatile organic compounds (Silvey and Roach, 1964). From these organisms, geosmin and 2-methyl isoborneol (MIB) have been identified as taste-and-odor-causing compounds (Gerber, 1979). While not toxic, these specific compounds reportedly impart threshold odors at concentrations in the range of 10 to 200 nanograms per liter (ng/l or $\mu g/m^3$) (AWWARF, *Advances in Taste-and-Odor Treatment and Control, 1995*, Table 5-2). To a degree, the compounds produced by cyanobacteria and actinomycetes can be degraded by bacteria. In 1965, Hoehn demonstrated the control of earthy-musty tastes by adding bacteria to Lake Hefner in Oklahoma City.

As algal and bacterial cells die and rupture (lyse), they release soluble internal cellular organic compounds into the water. As a result, from a taste-and-odor control standpoint, it is considered more desirable to *physically* remove microorganisms rather than rupture the cells with a disinfectant or oxidant during pretreatment. In other words, coagulation, sedimentation, and filtration are among the preferred methods for removal of the algal cells. Chemicals that cause cells to lyse and release their organic content may have the potential to further increase taste-and-odor problems.

LABORATORY CAPABILITIES FOR ASSESSING LAKE AND TREATED WATER QUALITY

Total Organic Carbon

Total organic carbon (TOC) is a composite measure of the organic content of water. While TOC may vary with season, nutrient-enriched lake waters may be particularly high in TOC during periods of stagnation, high water temperatures, and algal blooms. Measurements in 2004 indicated that Bloomington's treatment process was achieving approximately 57% TOC removal (Table 11-1). This exceeded the USEPA regulatory requirement of 25% TOC removal.

Repetitive analyses during the taste-and-odor episode indicated that the Bloomington plant influent exhibited a TOC concentration around 4.1 g/m^3. About half of this organic matter was being removed during lime softening and clarification. Despite this substantial TOC removal by entrainment in the sludge, the recycled water from the sludge storage lagoons, which had become far higher in geosmin during storage, was 20% lower in TOC than the lake influents. These results indicated that TOC and geosmin concentrations were not parallel. Moreover, as TOC was being degraded, geosmin was being formed in the sludge lagoons.

Nor were the compounds involved equally adsorbable. Filtration through the GAC-capped filters, while reducing geosmin concentrations by 50%, removed just 16% of the

TABLE 11-1 TOC Concentrations in Lake Bloomington and Finished Water

	TOC, g C/m^3		
Date	Lake Bloomington	Finished Water	% Removal
11/5/2004	3.4	1.4	59
12/1/2004	2.9	1.3	55

TOC applied to the filters. Finally, the TOC of the filtered, finished, and distributed waters averaged about 2.1 g/m^3.

Microscopic Analysis

On a weekly basis throughout the taste-and-odor episode, Bloomington's compound light microscope with epifluorescence (ultraviolet light) attachment was used to observe and photograph the algae, protozoans, fungi, and bacteria originating in the lake water as part of assessing their removal throughout the treatment process.

Tannins and Lignins

Bloomington had also recently acquired laboratory capabilities for measuring the combined *tannin and lignin* concentration of their source waters. This capability was utilized for the first time in the current evaluation of the lake water sources. The method is simple, rapid, and inexpensive and can be used in the field. It was hoped that this measure might serve as a useful adjunct to the time-consuming, direct measurement of TOC.

The first results of tannin/lignin analyses are shown in Table 11-2. Bloomington's lakes were sampled and aliquots drawn through a 0.22 μm neutron-track-etched polycarbonate membrane. These results indicated that there was little, if any, removal of tannins and lignins due to membrane filtration. This confirmed the expectation that this group of compounds are in true solution and that, if anything, this analysis might best relate to dissolved, rather than total, organic carbon.

Actinomycetes

Actinomycetes are filamentous, branching bacteria associated with earthy/musty tastes and odors. As organisms possessing properties intermediate between fungi and bacteria, actinomycetes have been referred to as *higher bacteria*. They are Gram-positive organisms that tend to grow slowly as branching filaments, resembling fungi, as their filamentous growth forms mycelial colonies. They were long regarded as fungi, as is reflected in their name: *aktino* (ray), *mykes* (mushroom or fungus).

Little published data is available on the source water abundance, physical removal, and inactivation of actinomycetes. However, the Kansas City (Missouri) Water Department has long monitored for these organisms in Missouri River water as part of their effort to avoid musty odors in their distributed water. A year of average monthly data from Kansas City is shown in Fig. 11-3. The results are based on bacterial plate counts and expressed as colony forming units per milliliter (cfu/ml).

TABLE 11-2 Tannin/Lignin in Bloomington Lake Waters

	Tannin/Lignin, g/m^3			
Sample Date	Evergreen Lake	Evergreen Lake, 0.22 μm filtered	Lake Bloomington	Lake Bloomington, 0.22 μm filtered
11/30/2004	1.2	–	0.7	–
12/1/2004	0.9	1.0	0.8	0.7

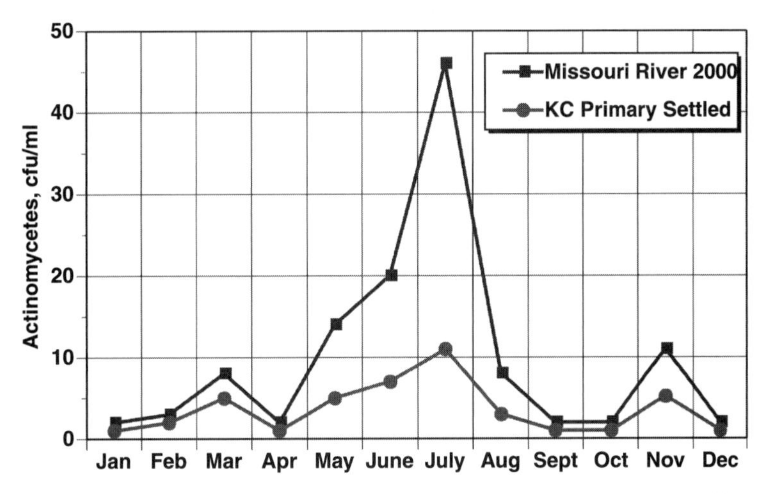

Figure 11-3 Actinomycetes in Missouri River water and their removal at Kansas City, Missouri.

As with most bacterial growth, including coliform, abundance is highest during warm weather months. This is why the situation in Bloomington, occurring when water temperatures were low, appeared to be an anomaly. Initially, it was speculated that a late autumn peak of actinomycetes might have resulted from the rain-induced influx of nutrients. It was later established that the rain event resulted in a cyanobacterial bloom of *oscillatoria* that caused the taste-and-odor event.

Geosmin and Methyl Isoborneol (MIB)

Analyses for the known odor-producing compounds *geosmin*, an alicyclic alcohol, and *2-methyl isoborneol (MIB)*, which is similar in structure to camphor, were conducted on samples collected in November 2004 (Table 11-3), in response to limited taste-and-odor complaints.

Subsequently, throughout December, with the continuing onset of odors, geosmin and MIB monitoring was increased markedly. The analytical results, illustrated in Fig. 11-4, confirm that readily detectable concentrations had developed in Evergreen Lake where the lake destratifier had been inoperative during the summer of 2004. In addition, an exceptionally wet autumn and winter had contributed to the influx of nutrient-rich tile drainage from the agricultural watershed.

Although TOC was substantially reduced, coagulation and lime softening did not appear to reduce odor compound concentrations significantly. Instead, softened and settled effluents sometimes appeared to exhibit even higher concentrations of geosmin and MIB than influents. Whether this increase is an artifact of high pH or is an inherent inaccuracy in sampling and measurement is unknown.

TABLE 11-3 Geosmin and MIB Concentrations in Lake, Finished, and Distributed Waters

Concentration, $\mu g/m^3$	Lake Bloomington	Finished Water	Distributed Water
Geosmin	10	7.2	11
2-Methyl isoborneol	0.7	5.3	5.5

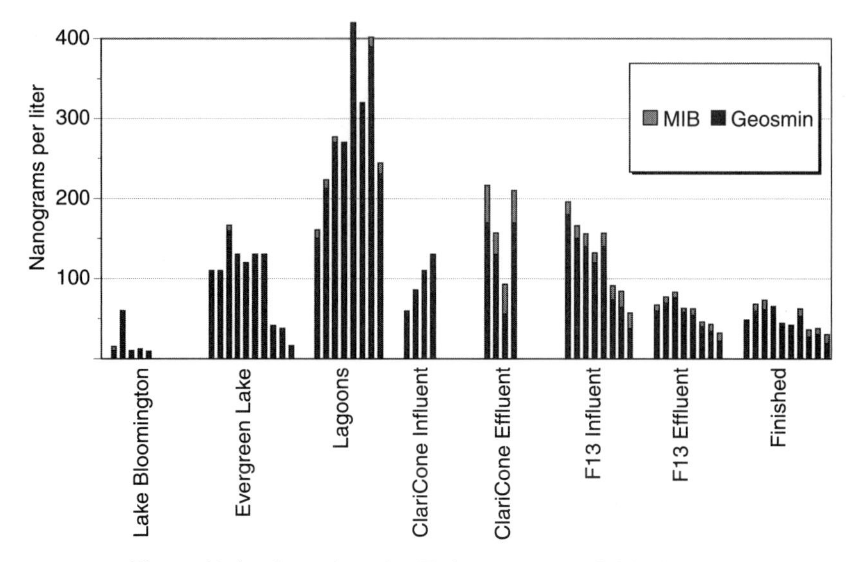

Figure 11-4 Geosmin and MIB from source to finished water.

However, within the plant, roughly half of the odorous compounds were removed during filtration through the GAC-capped sand filters. Earlier studies (June 2002) of Bloomington's filters had shown that both the GAC and filter sand support a thriving biological community. The diverse community of organisms on the filter media consisted primarily of attached bacterial cells. This evidence suggests that the odor-producing compounds may be partially biodegraded during their passage through the filter media. Accordingly, Bloomington operators slowed the passage of water through the filter media to allow extended contact time with the active microbial community. This was accomplished by extending filter runs and placing all operational filters into service to minimize average filter loading rates. Based on these observations, additional depths of contact media should be beneficial.

Oxygen and Hydrogen Sulfide

Analyses were conducted to determine whether there was oxygen depletion or the presence of reducing agents, such as hydrogen sulfide, in the lake waters. Both lake waters were fully oxygenated and hydrogen sulfide was measured at an insignificant 0.02 g/m^3. These results are consistent with lakes that have undergone the turbulence and destratification of their autumn turnover.

DESTRATIFICATION

Many Illinois water utilities utilizing lake water sources provide aeration and/or destratification to mitigate the adverse effects of anoxia and lake stratification. They seek reductions in taste-and-odor algal populations, elimination of zones of anoxia accompanied by the dissolution of iron and manganese from sediments, and reduced potential for subsequently forming trihalomethanes due to high organic concentrations. A prime example of such a destratifier unit that has been highly effective in controlling anoxia and algal blooms is the one operating in Lake Bloomington. Operated since 1997, this compact unit consists of a submerged pump that forces water through a Venturi injector, which, in turn, aspirates air into the water stream from an air line to the lake surface. Based on the success of the destratifier in Lake Bloomington, a similar destratifier was installed in Evergreen Lake (see Figs. 10-7 and 10-8).

To provide access to facilitate maintenance and repair, Bloomington maintenance staff has suggested that the destratifier pumps be relocated onshore. As matters stand, a diver is now needed for inspection and repair.

LAKE SOURCE WATER PROTECTION

Source water protection has been a major part of Bloomington's efforts to control drinking water quality, including taste and odor compounds. In addition to ongoing, long-term field studies of nutrient control in agricultural drainage plus an intensive program of lake water monitoring, Bloomington Water has implemented shoreline stabilization projects to protect the Lake Bloomington shoreline.

Limnological studies have determined the influence of rainfall and runoff on the residence time of water in each of Bloomington's water supply lakes. Lake Bloomington, a smaller lake in a larger watershed, provides a shorter in-lake residence time for equivalent precipitation. Heavy rains in the winter of 2004–2005 were estimated to have displaced twice the volume of Lake Bloomington (Fig. 11-5), whereas only 60% to 70% of Evergreen Lake was displaced. Due to nutrient flushing, conditions that favored high geosmin concentrations in Lake Bloomington may have diminished more rapidly than in Evergreen Lake.

Figure 11-5 Winter 2005 precipitation led to lake overflows.

The blending of the lake waters as plant influent is hampered by the configuration of the intake pumping facilities. Evergreen Lake has large pumps that supply a high proportion of the total treatment plant water demand. Therefore, fine-tuning of the blend between the two lake water supplies cannot be readily achieved. Accordingly, the design and installation of additional, smaller capacity pumps at Evergreen Lake would be beneficial to plant operation during periods when it is desirable to limit (blend) the input from that source.

ODOR CONTROL: OPERATING PROCEDURES AND PROCESSES

In summary, a number of operating procedures and treatment processes are routinely utilized to control organic materials and reduce tastes and odors. These include:

- Blending of Lake Bloomington and Evergreen Lake waters to minimize both nitrates and odors.
- Lake water quality and watershed management to limit the influx of algal nutrients.
- Operation of destratifiers to reduce algal growths and maintain aerobic conditions.
- Lake water pretreatment with cationic polymer for enhanced particle removals.
- Monitoring of organic matter (TOC, DOC) to optimize removals.
- In-plant powdered activated carbon (PAC) feed to increase short-term organic adsorption capacity.
- Granular activated carbon (GAC) filter caps on all 18 filters to, initially, adsorb dissolved organic matter and, subsequently, biodegrade labile organic compounds.
- Maintenance of persistent chloramine residuals in the distribution system to control bacterial regrowth and reduce chlorinous tastes.

In response to the November 2004 switch to Evergreen Lake water, Bloomington's powdered activated carbon (PAC) feeders were started at their maximum rates, feeding about 5 g/m^3. Initial problems with blockages of the PAC feeder ports were quickly overcome. However, after the initiation of feed, operators noted increases in softener effluent turbidities. This was particularly noticeable because, normally, water exiting the softeners has exceptionally low turbidity, commonly in the range of 0.2 to 0.7 ntu. As a result, even modest upsets are readily detected.

The turbidity increases were attributed to the PAC feed. In addition, there was concern over the passage of PAC fines though the filter as *particles of potential health significance.* This is because microorganisms that colonize PAC, a chemical reducing agent, may be protected from inactivation by disinfectant residuals. Figure 1.3k (see color insert) shows PAC fines embedded in microfloc along with fluorescing bacterial cells.

To assess the effectiveness of PAC dosages and available contact time in controlling odors, a laboratory test series was conducted on Evergreen Lake water using dosages of PAC ranging up to 20 g/m^3. After a 20 minute rapid mix, the PAC was allowed to settle overnight. The following day, comparable odors were detected in each of the six test beakers, indicating that little odor reduction had been accomplished. In addition, tannins/lignins were not measurably reduced by the PAC treatment.

FINISHED WATER TASTE AND ODOR MONITORING

Every few hours, Bloomington's plant operator-on-duty samples finished water from a continuously flowing tap in the operator's control laboratory. Analyses are conducted for disinfectant residuals, fluoride, alkalinity, pH, and other parameters, as necessary. An enhanced program of finished water odor detection has evolved as part of their effort to detect changing odor conditions before water reaches the distribution system.

When Bloomington operators become aware of increasing tastes and odors in the finished water, they notify laboratory personnel. Presently, analysis for known odor-producing compounds has shown that water withdrawn (recycled) from the lime softening sludge lagoon contributes the highest concentration of odorous materials to the influent water. This has led to an effort to decrease or eliminate the recovery of this water when odor levels are high.

OPERATIONAL ALTERNATIVES

Additional operational steps have now been adopted to control tastes and odors. These include:

- Increasing the dosages of cationic polymer at both lake intakes to improve particle, particularly algal cell, removals.
- Additions to the number of filters in operation to reduce filtration rates and increase empty bed contact time (EBCT) with GAC. Since the filters in the 1929 Annex provide a longer EBCT at lower filter loadings, these filters may be placed in service in lieu of those in the Main Building. However, reduced flow rates on all filters will provide longer contact times with the GAC caps.
- The cessation of prechlorination to avoid lysing of cells and confounding of tastes by chlorine.
- Initiation of a feed of an alternate (high specific activity) powdered activated carbon, at approximately $5\ g/m^3$. Moreover, the rarely-used PAC feed equipment is inspected more frequently to ensure it is fully operational when an odor emergency arises.
- Increased frequency of examination of lake water samples for organisms associated with tastes and odors plus laboratory testing for odorous organic compounds. To facilitate monitoring of organic compounds, it remains to be determined if there is a relationship between TOC, UV, and odor. Depending on appearance of microorganisms, consideration is being given to initiating the plate count for actinomycetes (taste-and-odor-causing bacteria).

EVALUATION OF POTENTIAL FOR MODIFIED OR ENHANCED TREATMENT PROCESS

Testing of alternative treatment processes is ongoing in Bloomington's laboratory. Special efforts are being made to take additional advantage of the microbial biodegradation of taste-and-odor-producing compounds on the biologically active GAC filters available in the plant. With initial testing showing that geosmin and MIB are currently being reduced by about 50% during filtration, laboratory studies are being directed at developing procedures for further enhancing *biodegradation*.

Other laboratory trials involve *aeration* for stripping of volatile compounds and the use of *chemical oxidants*, such as potassium permanganate, ozone, chlorine dioxide, and hydrogen peroxide (Fenton's reagent) to convert odorous compounds to odor-free constituents.

Still other biological and chemical efforts are directed at controlling odorous compounds directly at the lakes and lagoons. The lagoons, in particular, have been shown to be a major source of odorous compounds. Nearly half of the organic matter entering the plant is removed by softening and transferred with the softening sludge to the lagoons. Upon storage, it appears that this material spawns the development of odorous compounds. As a result, permission is being sought from the state regulatory agency to allow supernatant water from the storage lagoons to be discharged to a local stream.

Aeration

A sample of odorous water from the sludge lagoon was aerated overnight with a small air pump and diffuser stone. A second sample was used as a control. Both samples were kept in a temperature-controlled (cold) water bath. After this very extensive aeration period, the aerated sample was judged to smell fresher.

Owing to the in-house capability of the Bloomington Water Department laboratory (Fig. 11-6), the effectiveness of aeration was also evaluated by measuring the TOC of the lagoon water before and after aeration. Nominally, this is a measure of *purgeable* organic carbon. However, TOC is a relatively insensitive measure for this purpose because most of the compounds present in lake waters are relatively nonvolatile. Inexplicably, while prolonged aeration of the lagoon water resulted in a decrease in

Figure 11-6 Total organic and inorganic carbon analysis.

TABLE 11-4 TOC and Odor in Aerated Lagoon Water

Parameter	Aerated Lagoon	Lagoon (control)
TOC, g/m^3	3.9	3.0
Odor	Moderate	Strong

Figure 11-7 Incomplete odor reduction by aeration.

detectable odor, it was accompanied by an *increase* in TOC (Table 11-4). It is not known whether additional organic matter was introduced with the diffuser, container, pump, or laboratory air (Fig. 11-7).

Geosmin and MIB may be more sensitive measures of the removal of volatile compounds. However, in the absence of a very clear indication of the marked effectiveness of aeration in decreasing odors in the most odorous (lagoon) water, these costly analyses were not conducted.

Potassium Permanganate and Other Oxidizing Agents

Preliminary tests using lagoon water indicated that moderate dosages of potassium permanganate (1 to $2\,g/m^3$) either may have decreased or, perhaps, altered odors. These initial tests were followed by an evaluation of the change in oxidation-reduction potential (ORP) when titrating using a $1\,mg/ml$ permanganate solution. While its significance is not known, the ORP response curve increased and reached a plateau at a permanganate dose of about $1\,g/m^3$.

Preliminary tests with moderate dosages of oxidants, such as sodium chlorite, chlorine dioxide, hydrogen peroxide, and, for good measure, Fenton's reagent, failed to result in detectable odor decreases (Fig. 11-8).

Figure 11-8 Chemical oxidation with permanganate, chlorite, chlorine dioxide, and hydrogen peroxide.

Fenton's Reagent

Fenton's Reagent is a combination of iron and hydrogen peroxide that is reported to generate highly reactive *hydroxyl radicals*, which in turn remove odor and color. Reaching a pH of 6 in a two-liter Evergreen Lake water sample required the addition of a substantial quantity of ferric sulfate corresponding to a full-scale feed rate of 335 g/m^3. Thereafter, ferrous ion and hydrogen peroxide were applied for the oxidation of organic matter (Fig. 11-9). While laboratory measurements indicate that the TOC of the lake water was

Figure 11-9 Testing using Fenton's reagent.

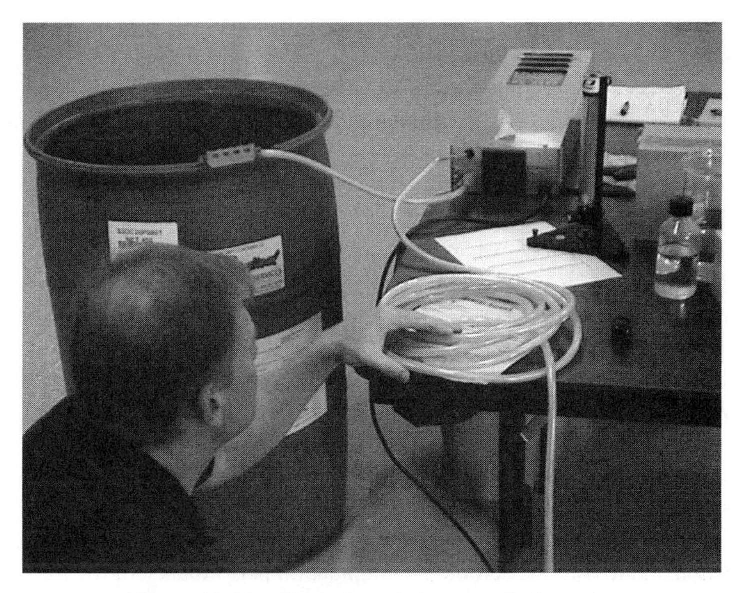

Figure 11-10 Ozonation of Evergreen Lake water.

reduced from 3.7 to 2.5 g/m^3, this treatment did not appear to achieve significant odor reduction. The precipitation of a substantial amount of hydrous ferric oxide may have helped remove some of the particulate fraction of the TOC.

Ozonation

Preparations were made for small-scale, laboratory studies using an ozonater powered by ultraviolet lamps. To facilitate this, efforts were made to prepare the apparatus for determining ozone output and the ozone demand of the lagoon and lake waters. If significant odor reduction could be demonstrated, some estimates were to be made of the capital and operating costs of full-scale ozonation (Fig. 11-10).

Biodegradation

Columns were also prepared for preliminary studies of *biodegradation* of odorous compounds on GAC removed from an operating filter. In addition to varying the EBCT and media content, it was proposed that some of Bloomington's filter media be sterilized (e.g., microwaved or washed) and utilized as a biologically inert control. The qualitative difference in odor reduction performance might then be attributable to the effect of aerobic biodegradation.

BIODEGRADATION: BLOOMINGTON'S WATER TREATMENT PLANT OPERATIONS MANUAL

In June 2002, a detailed study of Bloomington's filtration practice was undertaken as part of the effort to develop a comprehensive *Water Treatment Plant Operations Manual*. The

Filter Surveillance and Operation Manual (Chapter 5) outlines the design, installation, and configuration of Bloomington's 18 filters. It describes proper filter operation (per senior plant operators); USEPA filtration requirements, and monitoring of individual filter performances. The manual further details Bloomington's program of media replacement, contrasting the spent versus virgin GAC and the condition of the underlying sand. Of special interest are the micrographs showing the extensive development of microbial growth on both GAC and sand media (Figs. 4.3a, 5.23a–5.23e; see color insert).

Comparative operational data on the removal of total (primarily, dissolved) organic carbon by the virgin and in-service GAC is also presented in the manual. This illustrates the greatly reduced removal of TOC on GAC after one year of service and confirms that the adsorptive capacity of the virgin carbon is exhausted within months only to be partly supplanted by the removal of organic compounds by aerobic microbial respiration.

Recognizing the biological nature of the TOC removal process, a caution is included because of these early media studies. When a filter is removed from service, oxygen is found to decrease within hours. While DOC, ammonium, and nitrite ion concentrations in the filter pore water increase, attached microbial cell mass deteriorates under the anaerobic conditions. As a result, water quality degradation is observed upon filter start-up. To avoid this deterioration and restore the aerobic biodegradation performance of the filters, all idled filters are backwashed before being returned to service.

MICROSCOPIC EXAMINATION AND DESCRIPTION OF MICROGRAPHS

Microorganisms are believed to be the root cause of the earthy/musty tastes and odors emanating from Bloomington's water sources (Evergreen Lake, Lake Bloomington, and lagoon recycle which is approximately 10% of the total plant influent). Accordingly, samples were

Figure 11-11 Classification of microscopic particle abundance.

TABLE 11-5 Twait's Classification of Microscopic Particle Abundance (Scale: 1 [low] to 5 [high])

January 11, 2005	Bacteria	Algae	Diatoms	Inorganic
Mackinaw River	2	1	1	3
Evergreen Lake	3	3	3	3
Lake Bloomington	3	2	1–2	4
Lagoon recycle	3	4	3	4
Inline mixer	4	1	4	2
Clarifier effluent	3	1	0	4
Filter influent	2	2–3	1	3–4
Filter effluent	2	0	0	0
Finished water	2	0	0	1

taken of each of the source waters as well as throughout the treatment process in an effort to obtain some evidence of the removal, disintegration, or destruction of the odor-producing organisms during treatment.

As part of this evaluation, samples were collected, membrane-filtered through a 0.22 μm polycarbonate membrane, treated with live-dead fluorescent stain (Invitrogen Corporation), and prepared for microscopic examination by ultraviolet fluorescence microscopy. Bloomington's research-grade microscope facilitated organism identification and quantification (Fig. 11-11). In addition, it allowed high-quality micrographs to be prepared and archived for future reference.

During the taste-and-odor episode, hundreds of representative micrographs were prepared and a particle classification system devised to describe the relative abundance of the particles appearing in the micrographs. An example of the system is given in Table 11-5 for particles found in source waters and throughout the treatment process.

TASTE AND ODOR REMISSION AND PAC STUDIES

Bloomington's drinking water taste and odor complaints progressively diminished as geosmin and MIB levels decreased toward the end of March 2005. With the apparent end of this four-month episode, a consultant was retained to conduct laboratory studies of three alternative powdered activated carbons (PAC) to combat future taste-and-odor episodes. The results of this testing program indicated that higher dosages of the highest grade PAC (greatest specific surface area) applied for an extended contact period (two hours) would remove most of the radiolabeled MIB (surrogate odor compound) applied in the test program.

Although at a reduced sampling rate, Bloomington continued testing for geosmin and MIB. In June, 2005 elevated levels of geosmin again appeared in water withdrawn from Evergreen Lake. Since Lake Bloomington exhibited 20-fold less geosmin, plant influent was immediately converted to 100% Lake Bloomington. As a result, finished tap water tastes-and-odors remained minimal.

From Fig. 11-4, it is evident that geosmin was far more abundant than MIB in Bloomington's source waters. MIB was not even reported in many influent samples. Instead, MIB was more likely to be found in lime-treated (high pH) samples. This led to speculation that MIB may be formed as a chemical byproduct of the softening process. Even after softening, MIB was far less abundant than geosmin.

DOMINANT ORGANISMS IN EVERGREEN LAKE

The reappearance of geosmin in Evergreen Lake, but not in Lake Bloomington, provided a further opportunity to assess and contrast these waters to determine which (and how many) organisms might be associated with the production of odorous compounds since effective control of these undesirable species at the source might obviate the need for subsequent treatment to remove their objectionable by-products.

POTENTIAL FUTURE TRIALS

Titanium Dioxide Adsorption

Scottish researchers at Robert Gordon University (Aberdeen) have reported the successful removal of earthy-musty tastes and odors from surface waters using titanium dioxide and ultraviolet light (*AWWA Mainstream, 49:1:2005*).

Biodegradation

Biodegradation appears to be the most promising and effective method for destroying geosmin and 2-MIB. Even after the adsorption capacity of the GAC had been exhausted, passage through Bloomington's microbially colonized filters was found to remove about one-half of the geosmin in the filter influent. Bloomington's 2004–2005 episode might have been far more severe if microbial degradation of taste-and-odor-producing compounds had not occurred on the filter media.

Analytical Services Utilized in the Evaluation of Taste-and-Odor Removal

Geosmin and MIB analyses:	Environmental Health Laboratories ($350 per sample), 11 South Hill Street, South Bend, Indiana 46617
Primary analytical facility:	Bloomington Water Treatment Plant Laboratory, 25515 Waterside Way, Hudson, Illinois 61748
Microscopic analysis, micrographs:	Rick Twait
Nitrate analyses (lakes, tributaries):	Jill Mayes, Ron Stanley
Total organic and inorganic carbon:	Ron Stanley
Tannins/lignins:	H_2O'C Engineering, Columbia, Missouri
Nitrate analyses (finished water):	PDC Laboratories, Peoria, Illinois

REFERENCES

AWWA Mainstream (2005). 49: 1.

AWWARF (1995). Advances in Taste-and-Odor Treatment and Control. *AWWA*, Denver, CO.

Gerber, N. N. (1967). Geosmin, an Earthy-Smelling Substance Isolated from Actinomycetes. *Biotechnol. & Bioengrg.*, 9: 321.

Hoehn, R. C. (1965). Biological Methods for the Control of Tastes and Odors. *Southwest Water Works Journal*, 47: 3, 26.

Palmer, C. M. and Tarzwell, C. M. (1955). Algae of Importance in Water Supplies. *Public Works*, 86: 107.

Silvey, J. K. G. and Roach, A. W. (1964). Studies on Microbiotic Cycles in Surface Waters. *J. AWWA*, 56: 60.

Waksman, S. A. (1959). Actinomycetes; Nature, Occurrence and Activities, Williams and Wilkins Co., Baltimore, MD.

12

GAC ADSORPTION AND MICROBIAL DEGRADATION

REMOVAL OF GEOSMIN ON GAC

As part of the ongoing program of evaluation of alternatives for improving the removal of taste-and-odor-producing compounds, the Bloomington Water Department laboratory staff constructed and assembled test facilities to specifically evaluate the removal of geosmin on granular activated carbon. A major objective of the testing program was to determine the effectiveness of organism growth on the carbon in providing biologically mediated removal of odorous compounds after the initial adsorptive capacity of the GAC was exhausted.

Figure 12-1 shows the experimental apparatus consisting of four (19 mm ID, clear PVC) columns filled with GAC removed from Bloomington's operating filters. The columns were set up to observe their respective removals of geosmin, the odorous compound most commonly found in Bloomington's lake waters. It was hypothesized that the removal of geosmin would indicate the removal of all naturally occurring, odor-producing compounds.

Column 1 was filled with 0.6 m of fresh GAC (virgin Calgon WVG) to which filter influent was applied at a rate of 0.12 m/min. Consistent with the filter loading rates on Bloomington's new plant filters, this results in an empty bed contact time (EBCT) of five minutes. Barrels of the plant filter influent water were spiked with approximately 450 ng/l of geosmin (Fig. 12-2). A positive displacement pump was used to provide equal flow rates to all four columns.

Columns 2 and 3 were filled with 0.6 m of one-year-old and two-year-old GAC, respectively. This carbon had been removed from those operating plant filters in which the GAC had been in service for one and two years. Column 4 was filled with 1.2 m of two-year-old GAC to observe the effect of doubling the EBCT.

Water Treatment Plant Performance Evaluations and Operations. By John T. O'Connor, Tom O'Connor, and Rick Twait

Figure 12-1 Experimental set-up of GAC test columns.

Figure 12-2 Lake Bloomington influent spiked with geosmin.

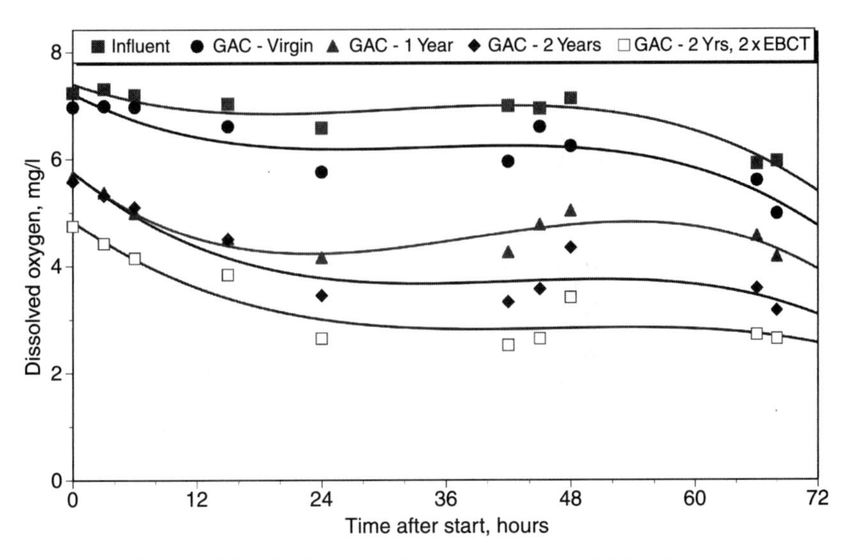

Figure 12-3 Depletion of dissolved oxygen in GAC columns.

Over a three-day period, column effluent samples were taken for analysis for geosmin and TOC. In addition, the filter effluents were monitored for dissolved oxygen and temperature.

Depletion of Dissolved Oxygen

Dissolved oxygen was measured to provide an index of microbial activity on the GAC media. However, even fresh (virgin) carbon was found to exert an oxygen demand of up to $1\ g/m^3$ after one day of operation (Fig. 12-3). One-year-old carbon consumed about $2.5\ g/m^3$ of the oxygen applied. Two-year-old carbon consumed an average of $3\ g/m^3$ at a five minute EBCT and about $4\ g/m^3$ when the EBCT was doubled.

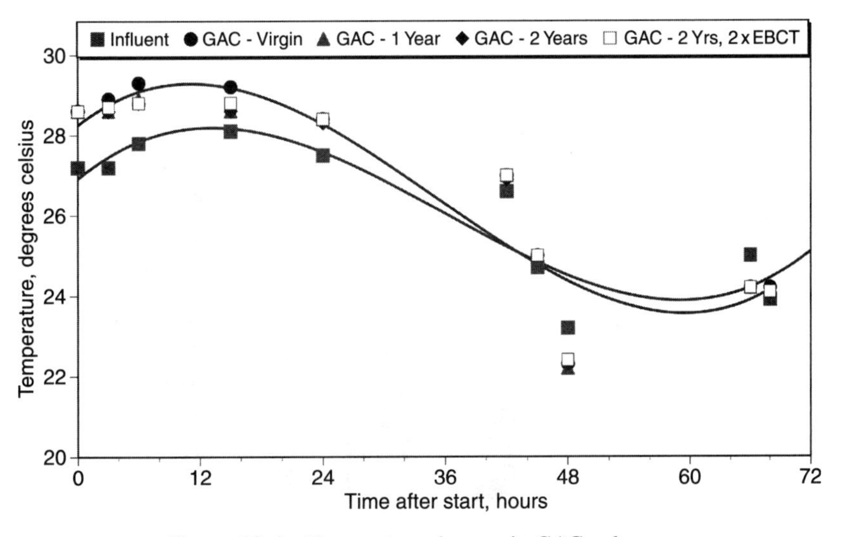

Figure 12-4 Temperature changes in GAC columns.

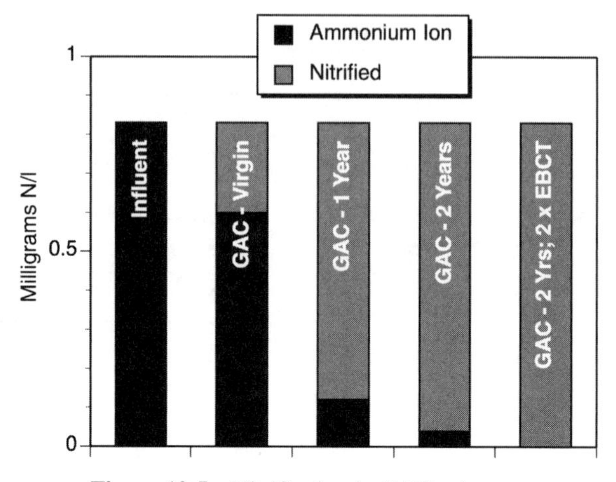

Figure 12-5 Nitrification in GAC columns.

The observed decreases in oxygen consumption in all columns after 36 hours appeared to be related to a marked decrease in influent water temperature. This temperature decrease influenced both oxygen solubility and the rate of microbial respiration (oxygen uptake). The two temperature curves shown in Fig. 12-4 illustrate the influent temperature variation and the effluent temperature of the column with the longest residence time.

Much of the oxygen depletion was related to the process of nitrification. After 68 hours of operation, analyses were conducted for ammonium ion in the influent and all column effluents. The results, shown in Fig. 12-5, revealed a progressive increase in nitrified effluent with both GAC service age and column residence time (EBCT).

The measured oxygen depletions in each column (Fig. 12-6) were compared with the calculated values of oxygen required for ammonium ion nitrification. The results indicated that most of the oxygen depletion was the result of the activity of the nitrifying organisms colonizing the media.

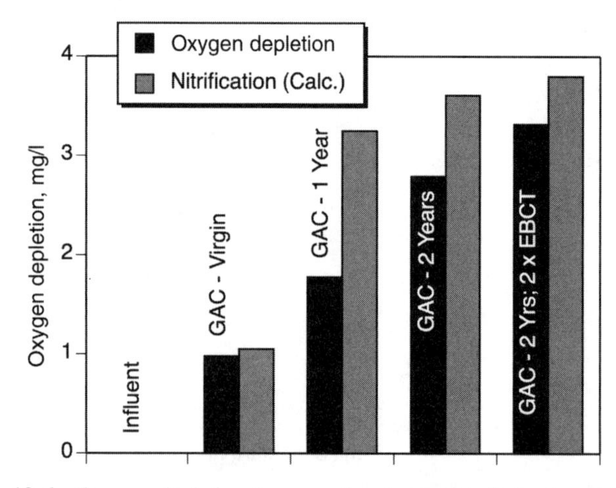

Figure 12-6 Oxygen depletion (measured and calculated) due to nitrification.

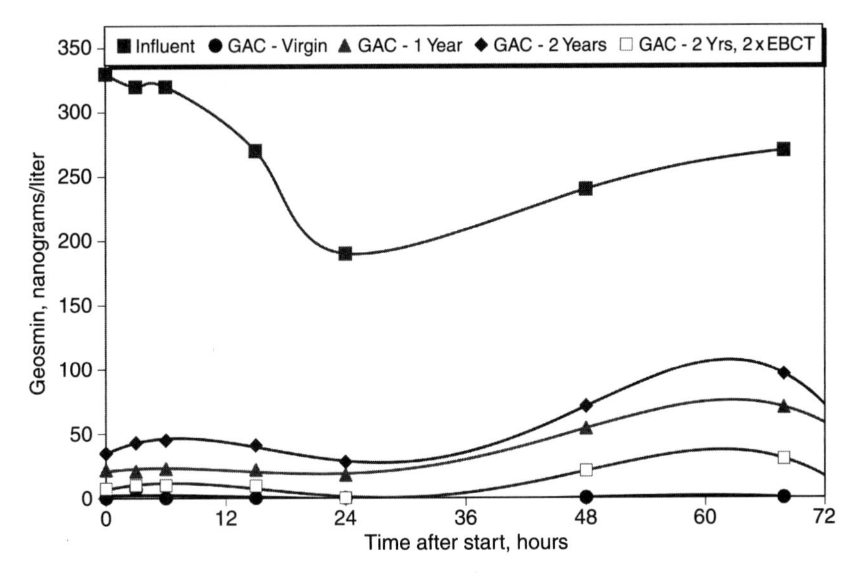

Figure 12-7 Removal of geosmin in GAC columns.

Removal of Geosmin

The influent geosmin concentration varied widely due, in part, to the significant amount of dilution required to reach the nanogram per liter range (Fig. 12-7). Moreover, some of the volatile geosmin may have been lost to the atmosphere. To better represent the results, the *percentage* removal of geosmin is plotted in Fig. 12-8.

Almost complete geosmin removal was achieved by adsorption on the virgin GAC. For the first 24 hours, even the one- and two-year-old carbon achieved removals in the 90% range. However, as temperatures declined and, presumably, as microbial activity decreased,

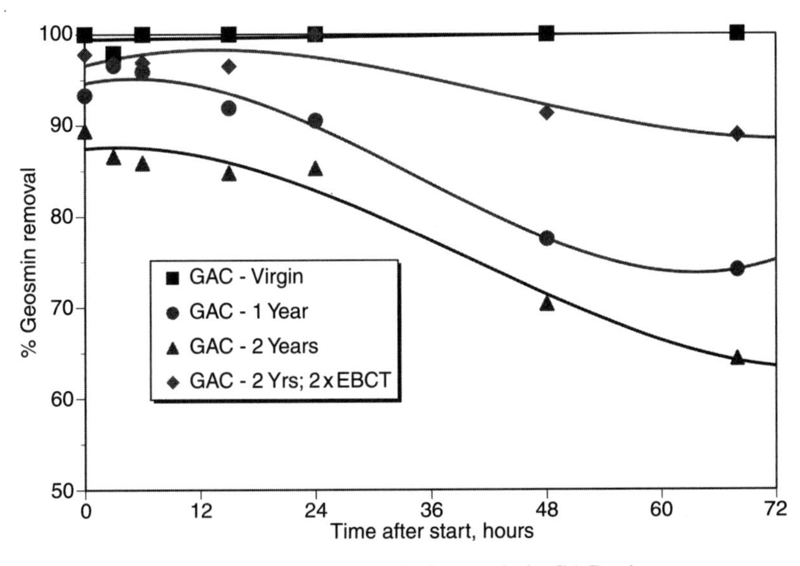

Figure 12-8 Percent removal of geosmin in GAC columns.

removals on the microbially colonized carbon declined. Even then, the columns removed 64% to 74% of the influent geosmin. Accordingly, with a doubling of the EBCT, the two-year-old GAC removed between 89% and 100% of the geosmin.

Since the adsorptive capacity of the GAC is commonly exhausted within months, the removals exhibited by the aged carbon appear to be primarily attributable to microbial assimilation.

Examination of GAC

To observe the accumulation of microorganisms on GAC with time of service in Bloomington's operating filters, aliquots of the virgin, one-year-old, and two-year-old carbons were extracted with 100 ml of demineralized, particle-free water (Fig. 12-9). After brief settling to remove the media, the extracts were decanted into 40 ml vials for subsequent measurement of turbidity and microscopic examination.

As shown in Fig. 12-10, turbidity visually illustrated the increased removal of solids from the aged filter media.

Figure 12-9 Extraction of accumulations from virgin and used GAC.

Figure 12-10 Turbidity of extractions from virgin and used GAC.

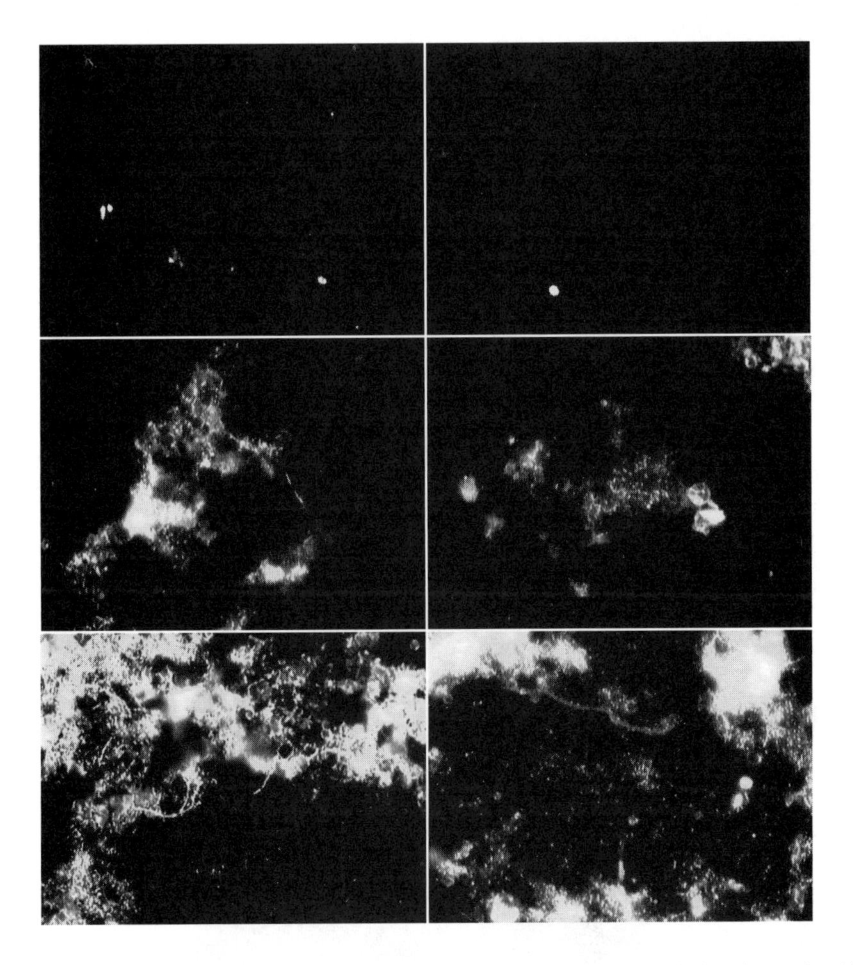

Figure 12-10a Comparative microscopic examination of GAC extracts with time in service (400x magnification). Row 1. Few fluorescing particles were observed on the extract from virgin GAC. Row 2. After one year, numerous bacterial cells, filaments, algae, diatoms, $CaCO_3$ crystals in gelatinous matrix. Row 3. After two years, abundance of unicellular and filamentous bacteria in gelatinous matrix; possible senescent filament. (See color insert.)

Microscopic Examination of GAC Extract

Ten milliliter portions of the extract were filtered through 0.2 μm polycarbonate membrane filters to retain bacterial cells and larger particles. These were then treated with a live-dead stain for fluorescence microscopy, rinsed, dried, mounted on a microscope slide, and viewed with an Olympus BX 60 light microscope under ultraviolet illumination. Micrographs of the particles and organisms recovered from virgin and aged carbons are shown on Figs. 5.23c, 5.23d, and 12.10a; see color insert.

REMOVAL OF GEOSMIN USING OZONATED AIR

The Bloomington Water Department utilized a small ultraviolet ozone generator for pilot-scale testing of the effectiveness of ozonated air in stripping or oxidizing taste-and-odor-producing compounds (Fig. 12-11).

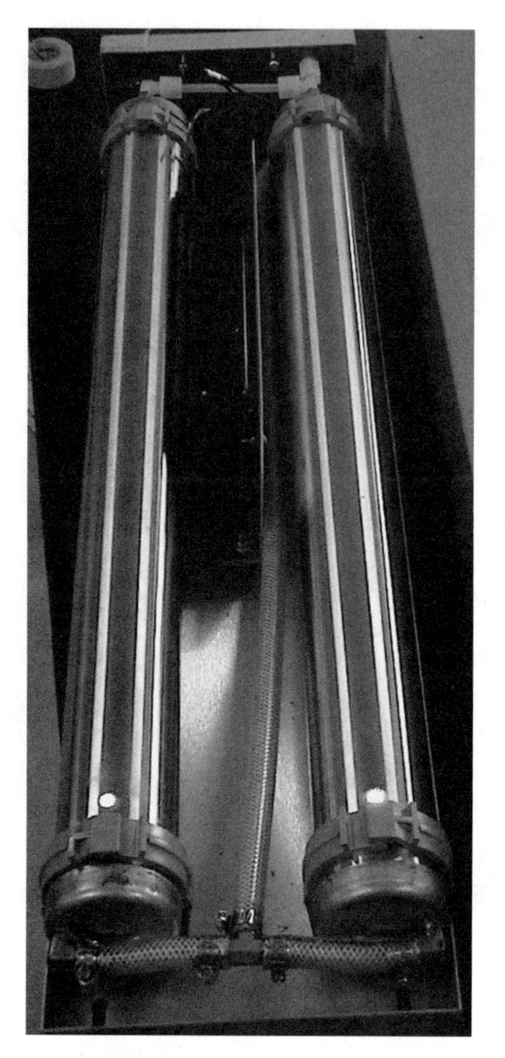

Figure 12-11 Ozone generator.

To evaluate the removal of geosmin, a barrel of Bloomington plant filter influent was spiked with approximately 450 $\mu g/m^3$ (nanograms per liter) and ozonated air was applied through diffusers located at the bottom of the barrel. While it was not determined what portion of the geosmin was removed by air stripping as opposed to oxidation by ozone, a 93% reduction in geosmin was accompanied by a reduction in the characteristic musty geosmin odor. However, the musty odor appeared to be replaced by a less intense odor that can be described as *ozonous*.

Follow-up comparative tests of air stripping, with and without ozonated air, were conducted to determine the relative effects of physical removal and oxidation by ozone. As shown in Fig. 12-12, after 6 hours, 25% of the geosmin and MIB were removed by air stripping alone while 83% of the combination was removed if ozone was applied.

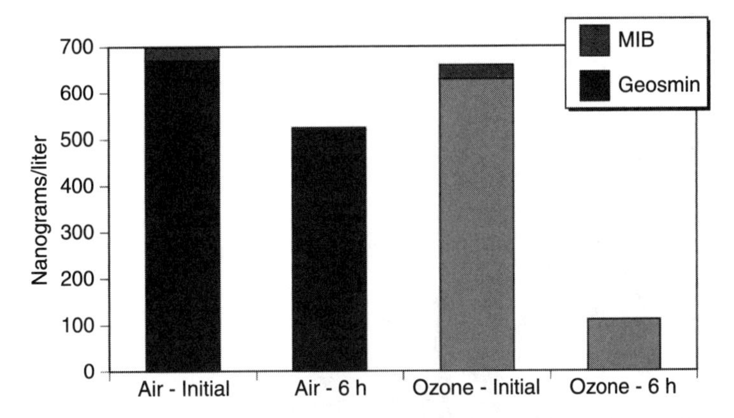

Figure 12-12 Removal of geosmin and MIB by aeration and ozonation.

LAKE WATER MONITORING FOR GEOSMIN AND MIB

The Bloomington Water Department continued to monitor its lake water sources and lagoons for geosmin and MIB as part of its long-term effort to control tastes and odors at the source. Data in Fig. 12-13 have been updated to October 2005.

These extensive data illustrate the notable success Bloomington's plant processes had in reducing the peak Evergreen Lake and lagoon influent geosmin and MIB concentrations. Often, concentrations were reduced from over 200 $\mu g/m^3$ to less than 50 $\mu g/m^3$ (nanograms per liter) in the finished water. Since, as shown in Figs. 12-14 and 12-15, much of this reduction was achieved within the GAC-capped filters, a more complete understanding of the relative effects of adsorption and microbial uptake on virgin and aged GAC was deemed vital to achieving improved operational control over periodic intrusions of naturally occurring odorous compounds.

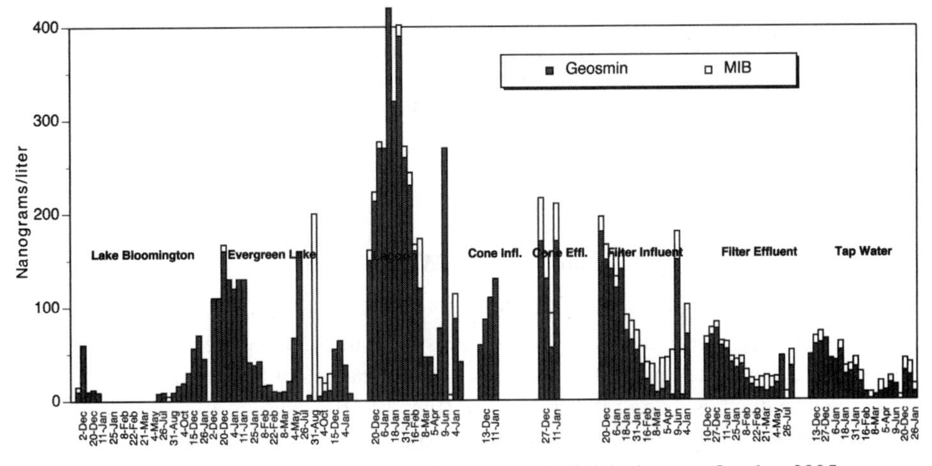

Figure 12-13 Geosmin and MIB from source to finished water, October 2005.

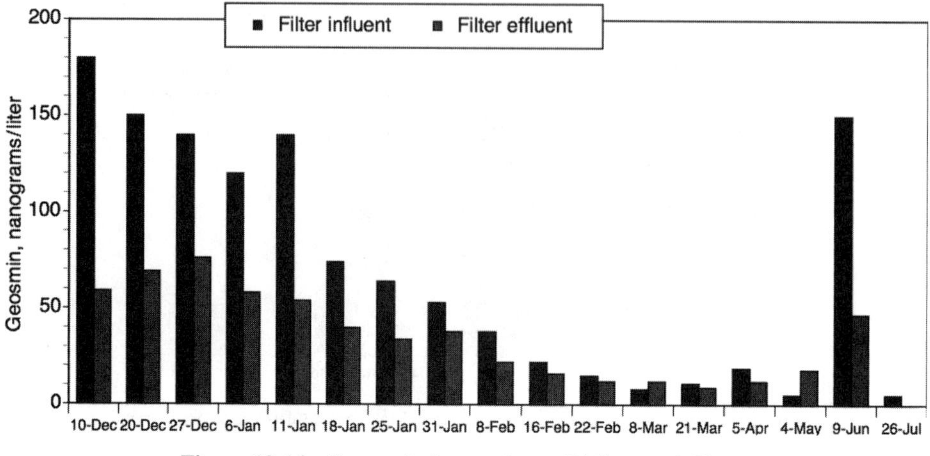

Figure 12-14 Removal of geosmin on GAC-capped filters.

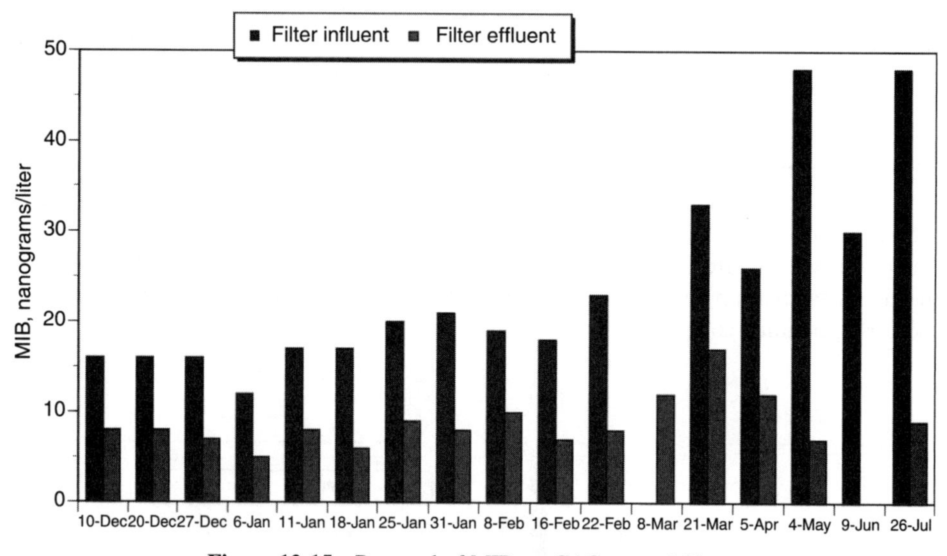

Figure 12-15 Removal of MIB on GAC-capped filters.

Geosmin and MIB as a Function of Lake Water Depth

An evaluation of the presence of geosmin and MIB as a function of lake water depth was conducted at Evergreen Lake. Analyses for geosmin and MIB revealed that these odorous compounds accumulate up to 20 times greater concentrations in the anoxic zone at and just above the lake benthos than in the epilimnion (Fig. 12-16). These results indicate that taste and odor problems may arise just after the lake experiences its autumn overturn. This is also the time when lake water temperatures begin their seasonal decline, diminishing biologically mediated treatment effectiveness.

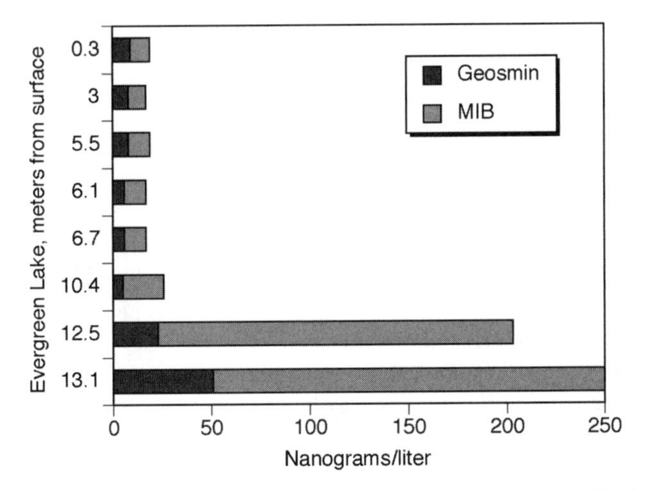

Figure 12-16 Geosmin and MIB as a function of lake water depth.

Depletion of Dissolved Oxygen in Operating Filters

To evaluate the extent of microbial activity several evaluations of oxygen depletion in Bloomington's operating filters have been conducted. In each case, oxygen was found to be depleted in the filter effluents. As indicated earlier, this varied with the age (time in service) and, therefore, microbial colonization of the GAC filter caps, the time since the most recent backwash, the temperature, and the amount of ammonium ion in the influent water.

For example, on June 14, 2005, with an influent water temperature of 24°C, oxygen depletions in five selected filters ranged from 0.35 to 1.72 g/m^3. Considering the short period of contact (<10 minutes) between the influent water and filter media, these results (Fig. 12-17) indicate rapid rates of dissolved oxygen uptake. These observations confirm

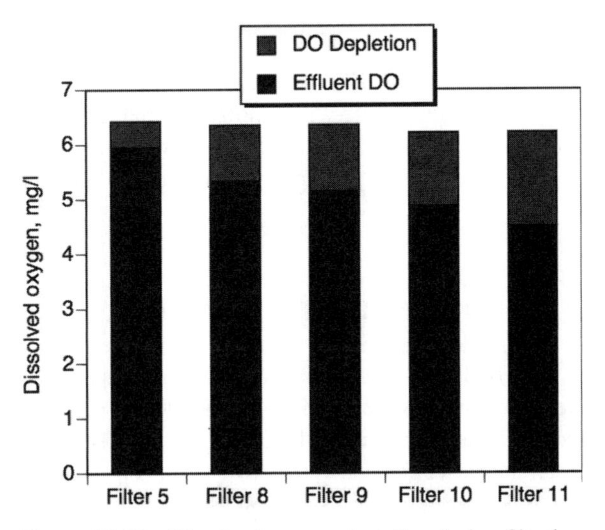

Figure 12-17 Dissolved oxygen depletion during filtration.

the expectation that temporary shutdown of individual filters may result in total depletion of oxygen and the onset of anaerobic conditions in portions of the filter bed. Without backwash, this could result in reducing conditions and septic odors on filter return-to-service.

Results and Conclusions

Column Studies The column studies clearly showed that, under flow conditions, dissolved oxygen was depleted in all GAC columns. The degree of depletion increased with the service age (microbial colonization) of the GAC and extended EBCT.

The effect of decreased column influent temperature was evident in that it decreased the rate of oxygen uptake and, presumably, the rate of microbial assimilation of odorous compounds. An important implication of this result is that the biologically mediated removal of geosmin and other odorous compounds may be seriously impaired when lake water temperatures are low.

Measurements of ammonium ion in the test water influent and GAC column effluents indicate that the observed oxygen depletion is largely due to nitrification. This may offer an opportunity for controlling dissolved oxygen depletion during filtration and water distribution by limiting ammonium ion in the source waters and filter influents.

Removal of Geosmin The removal of geosmin was virtually complete on the virgin GAC, indicating that the freshest carbon is the most effective in removing tastes and odors. However, there was also substantial removal of geosmin on the aged carbon. Removals ranged from 74% to 97% on the one-year-old GAC but decreased to the range of 64% to 89% on the two-year-old GAC.

Doubling the EBCT on the two-year-old GAC increased removals to the 89% to 100% range. This would imply that, as GAC ages, reduced hydraulic filter loadings would facilitate effective removal of odorous compounds.

Microscopic Examination Microscopic examination of GAC extracts confirmed little growth on the virgin carbon. Progressively more growth and gelatinous residue were rinsed from the carbons in service for one and two years. While there appeared to be a diversity of organisms present, most of the organisms observed were bacteria, both unicellular and filamentous.

Ozonation Although the results are very preliminary, a trial diffusion of ozonated air to geosmin-spiked filter influent reduced geosmin by 93%, seemingly eliminating its characteristic musty odor. However, the process appeared to produce another odor identified as *ozonous*.

FOLLOW-UP STUDY OF TEMPERATURE EFFECTS

A follow-up study was conducted to observe the effect of temperature on the removal of geosmin on one-year-old and two-year-old GAC. While all of the test GAC columns removed substantial amounts of geosmin, there was little difference in the performance of GAC that had been in service for one versus two years (Fig. 12-18).

Alternately, the temperature effect was very substantial. Far more complete geosmin removal was achieved when the water temperature was 25°C as opposed to 9°C. During

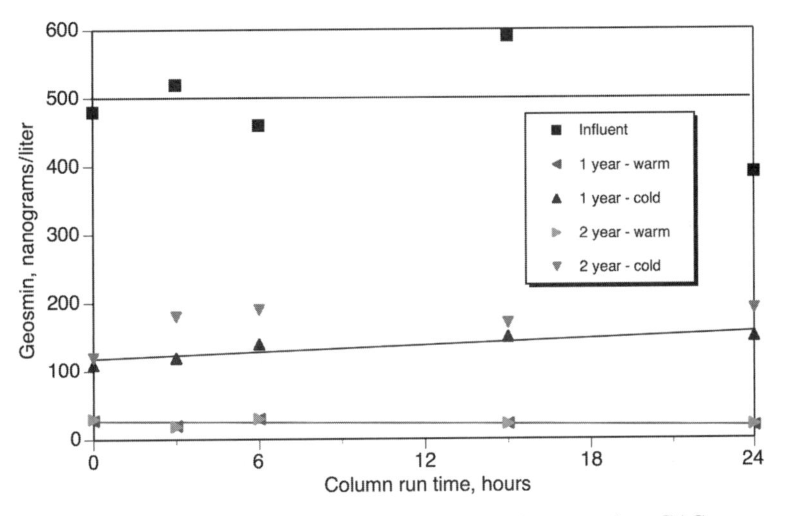

Figure 12-18 Effect of temperature on geosmin removal on GAC.

the winter months, as water temperatures decline further, a point is reached where microbial activity shuts down almost completely. This would leave any residual adsorptive capacity of aged GAC as the only mechanism for geosmin removal.

ULTRAVIOLET LIGHT PLUS HYDROGEN PEROXIDE

In November 2005, studies were conducted to evaluate the combined effectiveness of ultraviolet light and hydrogen peroxide (UV + H_2O_2) in promoting the oxidation of geosmin (Fig. 12-19). Softened and filtered water was collected and spiked with geosmin. This

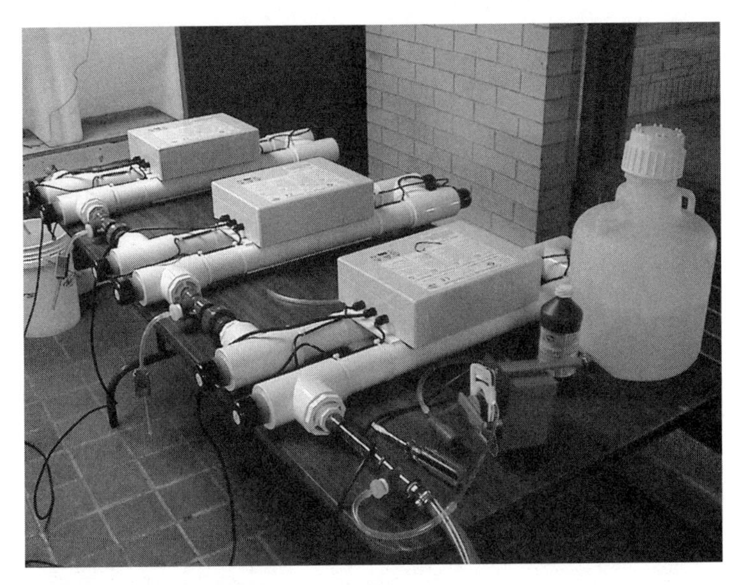

Figure 12-19 Ultraviolet light test apparatus.

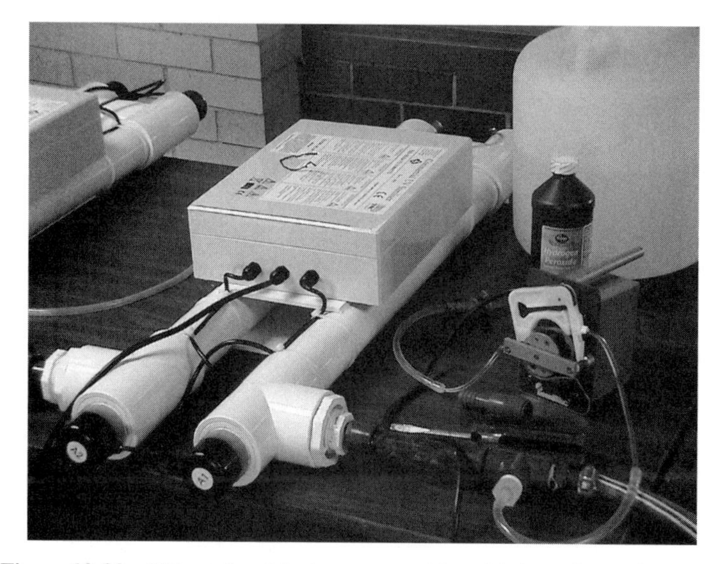

Figure 12-20 UV-catalyzed hydrogen peroxide oxidation of organic matter.

Figure 12-21 UV-peroxide test study.

was then pumped through a battery of three UV lamp pairs in series (Fig. 12-20). At first, there was no chemical feed, then hydrogen peroxide was fed by peristaltic pump through a small inline mixing device (Fig. 12-21). The initial tests were not successful in reducing geosmin concentrations. Subsequent runs with higher peroxide dosages (~35 mg/l) and extraordinarily long UV exposure times yielded complete removal of geosmin.

EARLY DETECTION AND CONTROL OF IMPENDING TASTE-AND-ODOR EPISODES AT EVERGREEN LAKE

Monitoring of Reservoirs for Geosmin and MIB

Taste-and-odor-producing compounds could evolve in either Evergreen Lake or Lake Bloomington. Accordingly, both sources are monitored regularly for geosmin and MIB as part of the water department's effort to detect trends in concentrations. In October 2005, monitoring with depth in Evergreen Lake showed that much of the geosmin and MIB was accumulating near the bottom sediments, possibly as a result of decomposition of algal cells and anoxic conditions. Although dilution with overlying water would be significant, subsequent mixing of the lake water during turnover would be expected to result in increases in the concentrations of these odorous compounds at the pumping station intake.

While costly, continued monitoring of known odorous compounds in the reservoirs should provide an early warning and help define seasonal patterns in the development of taste and odor problems.

Monitoring of Blue–Green Algal Blooms

Based on repetitive microscopic observations, blue–green algae are now believed to be the primary source of taste-and-odor-producing compounds in Evergreen Lake. Regular monitoring to observe populations and detect changes in the growth rates of blue–green algae are being used to provide an early warning of impending taste-and-odor episodes.

Bloomington has excellent microscopic facilities plus an experienced limnologist/microscopist/phycologist who is capable of identifying and quantifying blue–green algae. However, this is a specialized and labor-intensive effort. Numerous representative samples must be collected at the lake and prepared for microscopic examination. Thereafter, an experienced phycologist must correctly identify and provide, at least, a qualitative enumeration of the algae present.

A technological advance in digital imaging, flow cytometry, offers the potential for automating the identification and quantification of lake water algae. A flow camera microscope (www.fluidimaging.com) is offered for monitoring source water particles, including algae, ranging in size from 1 μm to 3 mm (Fig. 12-22). While the flow microscope system is costly, it may provide yet another means of obtaining an early warning of impending algal blooms. One response to this early warning might be the application of an algaecide capable of inhibiting the growth of the offending species.

Inhibition of Blue–Green Algal Blooms

A new method reputedly capable of specifically inhibiting the growth of blue–green algae has recently been described. Instead of copper salts that potentially have adverse effects on fish life and may adversely affect the entire algal community, sodium carbonate

Figure 12-22 Digital imaging flow cytometry.

peroxyhydrate (a blend of sodium carbonate and hydrogen peroxide) is reportedly capable of *selectively* inhibiting the growth of blue–green algae without disrupting the algal community as a whole. However, to be effective, it is recommended that this peroxide blend be applied at the onset of the algal bloom. Again, this would require close monitoring of the blue–green algal population.

IMPLICATIONS FOR PLANT OPERATIONAL CONTROL

The evaluation of GAC has provided a clearer understanding of the relative contribution of adsorption and microbial degradation to the removal of odor-producing compounds. Combined with the observed influences of temperature and residence time, operational strategies can be more readily defined for effective use of chemical feed and the capability of Bloomington's GAC caps.

Application of Powdered Activated Carbon

Consideration is being given to changing the point of application of powdered activated carbon from the plant site to the Evergreen Lake Pumping Station. Application at this point will result in a far longer contact period between the PAC and lake water. Although more costly, a higher specific activity PAC would be substituted for the low specific activity carbon presently applied at the plant's inline mixer. These modifications should result in an incremental increase in the removal of organic matter from the lake water. However, it is recognized that, by itself, PAC is not likely to control tastes and odors during an algal bloom.

Granular Activated Carbon

The extensive geosmin and MIB in-plant monitoring data obtained both before and throughout the extended 2004–2005 taste-and-odor episode has shown that Bloomington's GAC-capped filters provided the most effective removal of these odorous compounds.

Subsequently, the laboratory GAC column study results revealed that newer (fresh) GAC achieved almost total removal of high applied geosmin concentrations, whereas the GAC that had been in service for one and two years appeared to be largely dependent on microbial uptake for only partial geosmin removal.

From these results, it has been concluded that the best operational strategy for minimizing the passage of odorous compounds through the plant is to have the freshest possible GAC available at the time of an incipient odor challenge. Having determined the late autumn seasonality of the onset of recent odor problems, Bloomington Water has decided to modify the replacement schedule for GAC so that new virgin carbon will be placed in filters just before the autumn lake turnover.

Consideration is also being given to the replacement of the old filter bottoms in the Annex building with false-bottomed underdrains. This would eliminate the gravel layers, thereby allowing the filters to accommodate greater depths of sand and GAC media. Deeper GAC caps would also provide longer EBCT. These filter modifications should prove to be more economically feasible than alternatives that would both require new construction and markedly increase power consumption.

APPENDIX A

PROCEDURES FOR TOTAL BACTERIAL CELL COUNT BY EPIFLUORESCENCE MICROSCOPY

APPARATUS AND SUPPLIES

Acridine orange: 10 mg of acridine orange dissolved in 100 ml (0.1% W/V) of pH 7.2 phosphate buffer. Successful staining of bacteria can be achieved over a wide range of dilutions. Higher concentrations may yield brighter, more colorful microscopic images. Adjustment of lake, lime-softened, or filtered water pH by sample buffering does not appear to be necessary.

Membrane filters: 0.2 μm neutron track-etched polycarbonate membrane filters, black, 25 mm diameter, low fluorescence (available from Millipore, Nucleopore, or Poretics). Poretics and Millipore filter membranes were evaluated. Both provided excellent (black) backgrounds for epifluorescence microscopy.

Microscope: Olympus, Nikon, or Zeiss (Nikon E400 Microscope was utilized). Approximate cost (1997) was $11,000 with epifluorescent (410 nm) attachment plus phase contrast 10×, 40×, and 100× (oil immersion) objectives.

Reticle: 0.1 mm × 0.1 mm. Approximately 25,000 fields on membrane at 1,000× magnification. Approximately 1,000 fields on membrane at 200× magnification.

Immersion oil: Top of membrane—Type A; apply small drop to center of membrane. Top of cover slip—Nikon Immersion Oil or Cargille Type A.

Particle-free water: Prepare particle-free water by slowly filtering tap (plant finished) water through a 0.2 μm membrane filter. Prepare and count reagent blanks to ensure all extraneous particles have been removed. Glassware must be particle-free, but sterilization is not necessary.

Water Treatment Plant Performance Evaluations and Operations. By John T. O'Connor, Tom O'Connor, and Rick Twait

SAMPLING PROCEDURE

Bottles, 100 ml or larger, should be repeatedly rinsed with water to be sampled, then kept cool and in the dark to limit organism growth. No preservative is required if analyses are to be conducted within 12 hours. Sample mixing or blending is not necessary if particles remain dispersed.

Adjust sample volumes (1 to 50 ml) if they do not yield 200 to 500 total cell counts in 10 counting fields. Alternately, use of a consistent sample size (e.g., 5 ml) facilitates visual comparisons when viewing micrographs.

Membrane Filter Preparation

- After each use, rinse filter column with particle-free (0.2-μm membrane-filtered) plant finished water (Fig. A-1).
- Pipette sample (or particle-free water blank) into column and dilute to 10 ml with particle-free water to obtain uniform deposition of particles on membrane surface.
- Slowly filter diluted sample at low (<10 psig) vacuum to approximately 1 ml level.
- Add 1 ml ± acridine orange solution and allow to react for two minutes. Filter to near-dryness.
- Rinse unreacted acridine orange by adding 5 ml particle-free rinse water to column and filtering to dryness.

Slide Preparation

- Remove the membrane from the filter apparatus with forceps and air-dry for one minute or until there are no visible droplets or moisture. (Excess moisture on the membrane will form an emulsion with immersion oil and cause high background fluorescence.) Place the air-dried membrane on a microscope slide (Fig. A-2).
- Add a small drop of immersion oil to top (center) of membrane. Spread the oil with the placement of the cover slip and lightly flatten the cover slip atop the membrane to exclude bubbles. Do not place immersion oil under the membrane.
- Place slide on the microscope stage.

Microscopic Examination and Enumeration of Bacterial Cell Count

- Observe the slide under the 10× and 40× objectives to visualize particles larger than 3 μm and observe distribution of particles on membrane. Material on the slide must be evenly distributed for accurate counting (Fig. A-3).
- Apply immersion oil atop cover slide and observe distribution of bacterial cells using 100× (oil immersion) objective.
- If necessary, reduce the intensity of ultraviolet light using a neutral density filter (ND4). Intense light may cause fading of the cells due to desiccation.
- Randomly select ten (0.1 mm × 0.1 mm) fields and count all cells within each field. Record the number counted in each field on the data sheet. Observe the variation in the number of cells (or particles) counted per field. Calculate the average number of cells per field.
- An evaluation of the distribution of cells on the membrane may be estimated by calculating the *standard deviation* of the number of particles enumerated per field.

Determination of Number of Fields on Membrane Deposition Area

The measured diameter of the circle of deposition (wetted area) on the membranes used in these studies averaged 18.00 mm. This is not the internal area of the funnel, but the larger area on which particles have been deposited on the membrane. From these measurements, the following calculations were made:

Wetted deposition area $= \pi(9.00)^2 = 254.47\ \text{mm}^2$
Field area $= 0.1\ \text{mm} \times 0.1\ \text{mm} = 0.01\ \text{mm}^2$
Number of fields on deposition area $= 254.47\ \text{mm}^2/0.01\ \text{mm}^2 = 25{,}447$ fields

The number of fields will vary with the specific membrane filter apparatus (area of wetted deposition). Each new filter apparatus must be calibrated individually.

Calculation of Total Bacterial Cell Count

$$\text{Total Bacterial Cell Count/ml} = (\text{Average Count per Field}) \times (25{,}447\ \text{Fields})/\text{ml Sample}$$

Figure A-1 Vacuum flask and pump for membrane filtration.

Figure A-2 Membrane filter apparatus, forceps, microscope slide, timer, particle-free rinse water.

Figure A-3 Olympus microscope with epifluorescence attachment.

PRECISION OF EPIFLUORESCENCE TECHNIQUE

The precision of bacterial cell counting may be determined by repetitive counting of a number of slides from a sample (repeatability) or the reading of one slide a number of times from a sample (reproducibility). These numbers will vary with the operator. *ASTM D4455 (1990), Standard Test Method for Enumeration of Aquatic Bacteria by Epifluorescence Microscopy Counting Procedure*, has reported the repeatability and reproducibility of the method for high and low levels of bacterial cells, as follows:

Repeatability	Mean	Sample A: 6,200	Sample B: 8,600,000
Single Operator	Precision	1,400 (23%)	520,000 (6%)
Reproducibility	Mean	Sample A: 7,300	Sample B: 9,700,000
Single Operator	Precision	3,700 (51%)	890,000 (9%)

REFERENCES

APHA (1905). Standard Methods of Water Analysis: physical, chemical, microscopic and bacteriological methods of water examination (First Edition), American Public Health Association.

ASTM (1985). Enumeration of Aquatic Bacteria by Epifluorescence Microscopy Counting Procedure, D4455-85.

ASTM (1985). Simultaneous Enumeration of Total and Respiring Bacteria in Aquatic Systems by Microscopy, D4454-85.

ASTM (1988). Rapid Enumeration of Bacteria in Electronics-Grade Purified Water Systems by Direct Count Epifluorescence Microscopy, F1095-88.

Collins, V. G. and Kipling, C. (1957). The Enumeration of Waterborne Bacteria by a New Direct Count Method. *J. Appl. Bact.*, 20: 257-269.

Daley, R. J. (1979). Direct Epifluorescence Enumeration of Native Aquatic Bacteria: Uses, Limitations and Comparative Accuracy, Native Aquatic Bacteria: Enumeration, Activity and Ecology, ASTM STP 695, J. W. Costerton and R. R. Colwell, Eds., ASTM, 29-45.

Frankland, P. (1894). Micro-Organisms in Water-Their Significance, Identification and Removal. Longsmans, Green and Co., London.

Hobbie, J. E. et al. (1977). Use of Nuclepore Filters for Counting Bacteria by Fluorescence Microscopy. *Appl. Environ. Micro.*, 33: 1225-1228.

Mittelman, M. W. et al. (1983). Epifluorescence Microscopy, A Rapid Method for Enumerating Viable and Nonviable Bacteria in Ultrapure-Water Systems. *Microcontam.*, 1: 32, 52.

Mittelman, M. W. et al. (1985). Rapid Enumeration of Bacteria in Purified Water Systems. *Med. Dev. and Diag. Ind.*, 7: 144.

Newell, S. Y. et al. (1986). Direct Microscopy of Natural Assemblages, Bacteria in Nature, Vol. 2: Methods and Special Applications in Bacterial Ecology. J. S. Poindexter and E. R. Leadbetter, Eds., Plenum Press, New York.

O'Connor, J. T. et al. (1985). Chemical and Microbiological Evaluations of Drinking Water Systems in Missouri. *Proc. AWWA Annual Conf.*, Washington, D.C.

O'Connor, J. T. (1990). An Assessment of the Use of Direct Microscopic Counts in Evaluating Drinking Water Treatment Processes, ASTM Special Technical Publication 1102: Monitoring Water in the 1990's: Meeting New Challenges.

O'Connor, J. T. and O'Connor, T. L. (2002). Control of Microorganisms in Drinking Water, Chapter 8: Rapid Sand Filtration. American Society of Civil Engineers, ISBN 0-7844-00635-9.

Palmer, C. M. and Tarzwell, C. M. (1955). Algae of Importance in Water Supplies. *Public Works*, 86: 107.

Pettipher, G. L. and Rodriques, U. M. (1982). Rapid Enumeration of Microorganisms in Foods by the Direct Epifluorescent Filter Technique. *Appl. Environ. Microbiol.*, 44: 809.

Pettipher, G. L. (1983). The Direct Epifluorescent Filter Technique for the Rapid Enumeration of Microorganisms. Research Studies Press Ltd, Letchworth, Herts. SG6 3B3, England.

Silvey, J. K. G. and Roach, A. W. (1964). Studies on Microbiotic Cycles in Surface Waters. *J. AWWA*, 56: 60.

Stevenson, L. H. (1978). A Case for Bacterial Dormancy in Aquatic Systems. *Microbial Ecology*, 4: 127.

Syrotynski, S. (1971). Microscopic Water Quality and Filtration Efficiency. *J. AWWA*, 63: 237–245.

Waksman, S. A. (1959). Actinomycetes; Nature, Occurrence and Activities, Williams and Wilkins Co., Baltimore, MD.

Whipple, G. C. (1899). Microscopy of Drinking Water, John Wiley, New York City.

APPENDIX B

POTENTIAL STUDIES INVOLVING MICROSCOPIC PARTICLE ANALYSIS: PARTICLE IDENTIFICATION, ENUMERATION, AND SIZING

Based on experiences with plant process performance evaluations, the following list of potential studies involving microscopic particle identification, enumeration, and sizing was compiled.

Source Waters

Seasonal Changes in Source Water Algal and Bacterial Populations

Effect of Lake Destratification and Overturn on Particle Type, Size, and Number

Monitoring for Presence of Blue-Green Algae

Plant Influent

Evaluation of Effectiveness of Inorganic Coagulant and Polyelectrolyte Additions on Initial Particle Agglomeration

Lime Treatment, Coagulant Dosage, Clarification

Evaluation of Entrainment of Source Water Particles in Calcium Carbonate and Magnesium Hydroxide Precipitates

Removal of Particles Produced by Softening as a Function of Coagulant and Polymer Dosages

Comparison of Performance of Upflow Slurry Blanket Contact Clarifier Units of Different Design

Effect of Hydraulic (Surface) Loading Rate on Clarifier Performance

Recarbonation Basin Effluent

Effect of Recarbonation on Rate of Dissolution of Residual Calcium Carbonate

Effectiveness of Polyphosphate Addition on Stabilization of Calcium Carbonate

Water Treatment Plant Performance Evaluations and Operations. By John T. O'Connor, Tom O'Connor, and Rick Twait

Granular Activated Carbon Adsorption/Sand Filtration

Observation of Individual Filter Particle Removal Performances through Backwash Cycles: Effect of Temperature and Alternate Filter Washing Protocols

Assessment of Benefits of Flow Ramping on Filter Return-to-Service

Presence of Carbon Fines in Filter Effluent Resulting from GAC Abrasion

Calcium Carbonate Post-Precipitation in Finished Water on Prolonged Storage

Monitoring Infrequently Observed Particles in Finished Water: Nematodes, Rust Particles, Paint Chips, Oil Droplets, Sloughed Organic Debris

Overall Treatment Plant Performance

Evaluation of Effects of Seasonal Influent Water Temperature Changes on Overall Treatment Plant Particle and Organism Removal Efficiencies

ALTERNATIVE MEASUREMENTS FOR OPERATIONAL CONTROL OF WATER TREATMENT PROCESSES

Turbidity (Optical)

Operational Control Using Continuous Flow Turbidimeters on Filters

Laboratory-Grade Turbidimeter as Reference Standard

Gravimetric (Mass)

Solids from Softener/Clarifier Grit Removal and Slurry Blanket Blowdown; Filter Backwash

Electronic Particle Counter (Number and Size Classification)

Correlation of Electronic Particle Counts with Turbidity Measurements

Comparison of Numbers of Particles Indicated by Electronic Particle Counter with Numbers Observed and Enumerated on 3 μm Membrane Filters

Microscope (Light)

Enumeration, Sizing, and Identification of Particles >3 μm (Referee Method)

Enumeration and Sizing of Carbon Fines in Filter Effluent

Microscope (Epifluorescent)

Evaluation of Efficiency of Removal of 1 μm Sized Biotic Particles (Bacteria) for Confirmation of Removal of Particles of Potential Health Significance

Evaluation of Alternate Coagulants and Determination of Required Coagulant Dosages as a Function of Water Temperature

UNITS OF MEASUREMENT

As an aid to familiarizing water plant personnel with modern scientific terminology, most of the units used in the current volume are based on the International System of Units (SI), now the world's most widely used system of units. Some of the most common conversions were, as follows:

Length: 1 foot, ft. = 0.305 meter, m = 305 millimeters, mm
Area: 1 sq. foot, ft^2 = 0.093 sq. meter, m^2
Weight: 1 pound, lb. = 454 grams, g = 0.454 kilogram, kg
Volume: 1 gallon, gal. = 3.785 liter, l = 0.003785 m^2
Flow: 1 gal/minute, gpm = 0.227 cubic meter/hour, m^3/h
Flow: 1 mgd = 157 m^3/h
Velocity: 1 gpm/ft^2 = 2.444 meters/hour, m/h
Pressure: 1 psi = 6895 Pascal, Pa
Power: 1 HP = 746 Watts

APPENDIX C

DEVELOPMENT OF OPERATOR GUIDELINES

Half of all U.S. water plant operators are expected to retire in the next five years. This has alerted water utility management to the need for succession planning for a changing workforce. In implementing an *early retirement incentive* program in 2008, it appears that the City of Bloomington will be in the vanguard of this projected transition.

Fortunately for Bloomington's water utility, over the past decade, there has been a modest, but systematic, program of evaluating treatment processes and reviewing operational procedures. Even so, the rapid transition in experienced operational and maintenance staff threatens to place a particular burden on utility management. A series of managerial issues must be addressed in the near future, including:

Succession—Who will be operating the utility in the future? The selection of the most capable, qualified, and reliable operating personnel is often viewed as the primary function of any water utility administrator.

Recruitment—A shrinking pool of experienced, and certified, operating personnel makes recruitment a growing challenge that may require modification of traditional limitations on hiring (e.g., residency requirements). Also, increased competition for the most capable, certified operators may influence salary requirements.

On-the-Job Training—Will an accelerated in-house, system-specific training program be required to integrate new personnel? At what cost and who will provide the training?

Information Transfer—How well have *standard operating procedures* been documented and how can they be effectively communicated to new personnel?

Continually under development, the *Bloomington Water Treatment Plant Operations Manual* has been created to serve as an introduction to treatment plant configuration

Water Treatment Plant Performance Evaluations and Operations. By John T. O'Connor, Tom O'Connor, and Rick Twait

and operation for newly recruited plant personnel. It begins with the following series of abbreviated, illustrated, *guideline documents* prepared to supplement the *Plant Operations Manual* and facilitate utilization by new personnel.

1. Operator-on-Duty: Responsibilities
2. Operator's Laboratory: From Analysis to Database
3. Operator-on-Duty: Making the Rounds
4. The Lime Delivery System: From Dry Storage to Slakers to ClariCones
5. Recarbonation: From Liquid CO_2 to Gas to Solution
6. Filtration: Description and Operation

OPERATOR-ON-DUTY: RESPONSIBILITIES

The operator-on-duty monitors and controls the water production and purification process throughout the treatment plant (Fig. C-1). This operator also initiates the appropriate responses to all alarms and emergencies, such as power outages and equipment failures. Upon restoration of power after an outage, this operator must ensure that the plant pumping and treatment units have returned to proper operation.

Upon changing shifts and assuming control, the incoming operator-on-duty will note:

- Plant discharge pressures (about 43 psi on the high service pump discharge line pressure gauge)
- The plant water production rate (and adjust output, when necessary to meet system demand)
- The level of finished water available in online (downtown) storage facilities
- The intake flow, production, and system demand changes that have occurred over the past shift

Figure C-1 Operator's Control Room and Monitors.

Every two hours, the operator-on-duty *makes the rounds* to directly observe the operational condition of the treatment plant unit processes. This includes observation of the new and old filters, filter galleries, chemical feed systems (chlorine, ammonia, fluoride, ferric sulfate, lime, polyphosphate, anionic and cationic polymers, etc.), compressors, softening basins, and recarbonation system. During the rounds, samples are collected for analysis in the operator's laboratory.

A change in pumping rate due to variation in system demand requires resetting of chemical feed rates to maintain proper dosages. Chlorine and ammonia dosages are carefully monitored following any change in plant production. Since each of the lime softening units is generally operated near an optimal steady-state flow, basins may be added to or removed from service to match demand. This requires the addition or termination of the lime slurry flow to that basin.

In addition to observing filter head loss and turbidity, the operator-on-duty records the number of hours each of the 18 filters has been in operation. As a minimum, after 48 hours of service, a filter is backwashed. This is typically done manually so that filter operations (surface wash, release of air, media boils, backwash water appearance, etc.) can be observed. During filter production, the turbidity of the filtrate is continuously monitored by online turbidimeters and the results recorded to a SCADA system to provide a permanent record of filter performance. If any turbidity exceeds 0.3 ntu, the operator will manually collect a sample from that filter and analyze for both unacidified and acidified turbidity using the ratio turbidimeter in the laboratory. This will allow for a more definite response to the IEPA, the regulatory authority, if a defined turbidity exceedance actually occurs. A filter that continues to produce a filtrate turbidity >0.3 ntu will be shut down.

OPERATOR'S LABORATORY: FROM ANALYSIS TO DATABASE

The operator-on-duty ensures the proper operation of the plant processes by collecting samples every two hours as part of making the rounds of treatment units. Samples are collected of the settled effluents of those softening units (ClariCones) in operation (*Cone 1*, *Cone 2*, *Cone 3*, and *Cone 4*); of the effluents of the recarbonation basins (*CO_2-East* and *CO_2-West*) and of the filtered and disinfected (finished) water (Lab *Tap 01*).

The turbidity of each sample is measured using a Hach 2100N ratio turbidimeter, which is routinely calibrated by the plant control laboratory. Both *unacidified* and *acidified* samples (Fig. C-2) may be measured to determine how much of the turbidity is acid-soluble (primarily

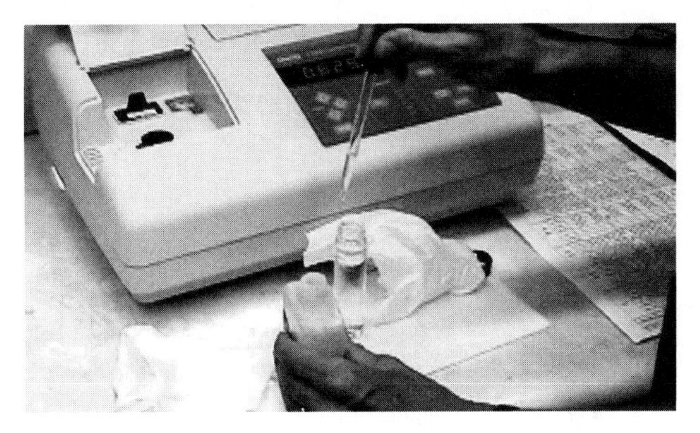

Figure C-2 Turbidity analysis—with and without acidification.

due to calcium carbonate). Acidified turbidities are measured in every instance where finished water samples exceed 0.3 ntu, which is, nominally, the EPA limit on filtered waters.

pH (a unit of hydrogen ion concentration) is a primary control parameter for the lime softening process. Although seasonally adjusted, the pH of the softener effluents may be maintained in the range of 10.5 to 10.7. Higher values indicate the likelihood of lime *overfeed* to the softener leading to subsequent post-precipitation of calcium carbonate during recarbonation; lower values may indicate failure of the lime slurry feed flow to the unit.

After recarbonation (stabilization), the pH is generally in the range of 9.5 to 9.7 and the water should remain clear as it flows to the filters. With the addition of chlorine following filtration, the finished water pH further decreases to 8.8 to 9.0.

Softening results in the removal of *alkalinity*. However, maintenance of adequate alkalinity, for its acid buffering capacity, is a concern with respect to corrosion of household plumbing. The total (methyl orange) alkalinity of the finished water should not fall below 18 mg/l as calcium carbonate equivalent.

Once per shift, or more frequently if changes are made to the plant flow, the finished water is tested for chlorine residuals. If free chlorine is greater than 0.3 mg/l, insufficient ammonia is being added to convert the chlorine to a combined (chloramine) residual. Total chlorine (primarily, chloramine) is generally maintained in the range of 3.5 to 3.8 mg/l as chlorine; fluoride, 0.9 to 1.2 mg/l; and total hardness >80 mg/l as $CaCO_3$ equivalent.

Plant operational data is entered into the database at the end of each shift (Fig. C-3).

Use Per Million Gallons Treated, in kWh; lb.			
Energy:	1,100	Lime (CaO):	1,294
CO_2:	226	$Fe_2(SO_4)_3$:	91
Chlorine:	44	Fluoride:	43
Ammonia:	12	Phosphate:	4

Figure C-3 Logging of plant performance data.

OPERATOR-ON-DUTY: MAKING THE ROUNDS

Every two hours, the operator-on-duty makes the rounds to directly observe the operational condition of the treatment plant unit processes. This includes observation of the new and old filter galleries, chemical feed systems (chlorine, ammonia, fluoride, ferric sulfate, lime,

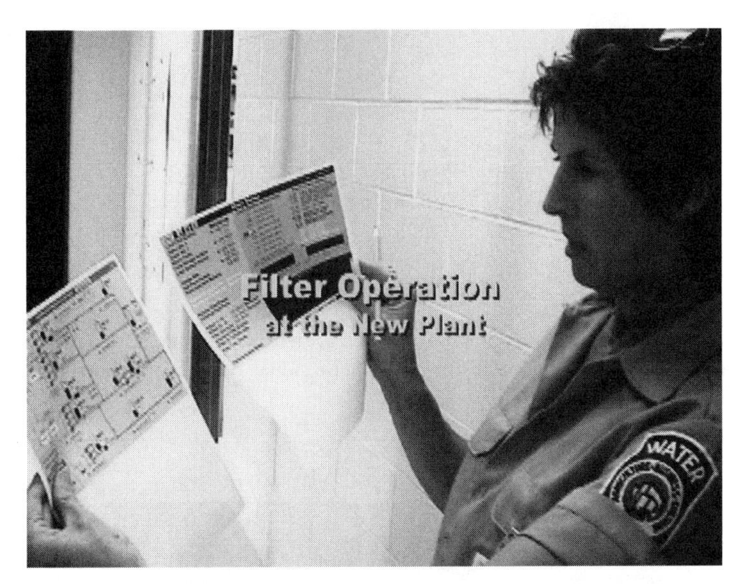

Figure C-4 Inspecting a printout of plant operating conditions on the way to the filter gallery, the operator observes flow rates, looks for flooding, listens for unusual sounds, and notes which filters will require backwash during her shift.

Figure C-5 Plant influent flows are balanced against discharge (production) rates to distribution system and storage.

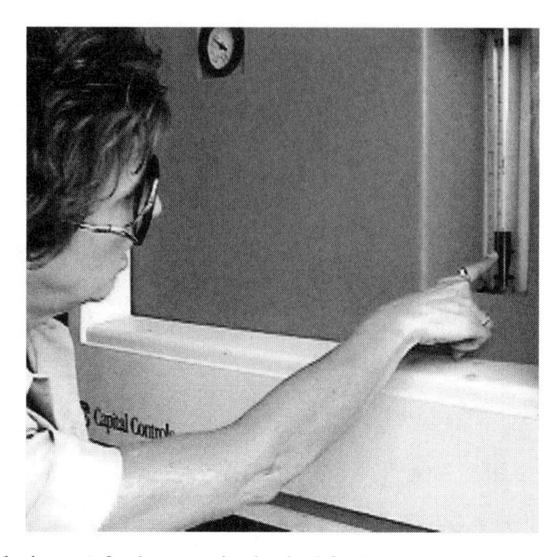

Figure C-6 The chlorine gas feed system is checked for leaks, odors, icing, and a satisfactory feed rate that will ensure a combined chlorine, finished water residual. Subsequently, one part of ammonia is fed per 3.5 parts of chlorine.

Figure C-7 Fluoride (1 mg/l ±) and ferric sulfate (91 lbs./mil. gal) feed pump operation and settings are checked.

Figure C-8 Compressors, vital to operation of valves, are checked for performance, pressure, unusual sounds.

Figure C-9 Bloomington's modern lime feed system is complex, but still prone to blockage and malfunction. Stored above two batch slakers, blockage may occur when powdered lime is metered from the hopper into the slaker tank.

Figure C-10 Control panels indicate each batch phase of lime and water addition, mixing time, temperature, and discharge of slaked lime to continuously-mixed slurry storage tanks.

Figure C-11 Slurry storage tanks, flexible hose, couplings, and lime slurry pumps are checked for leaks and incipient ruptures.

Figure C-12 Special precautions are taken when inspecting chlorine facilities for room ventilation, cylinders in service, rate of cylinder weight loss (feed rate), gas pressure, automatic changeover, alarms, and evidence of leakage (odor).

Figure C-13 Rotameter gas feed rates must be adjusted whenever plant flow rates are changed to match system demand.

Figure C-14 Carbon dioxide is fed to east and west recarbonation basins to reduce pH of lime-softened water.

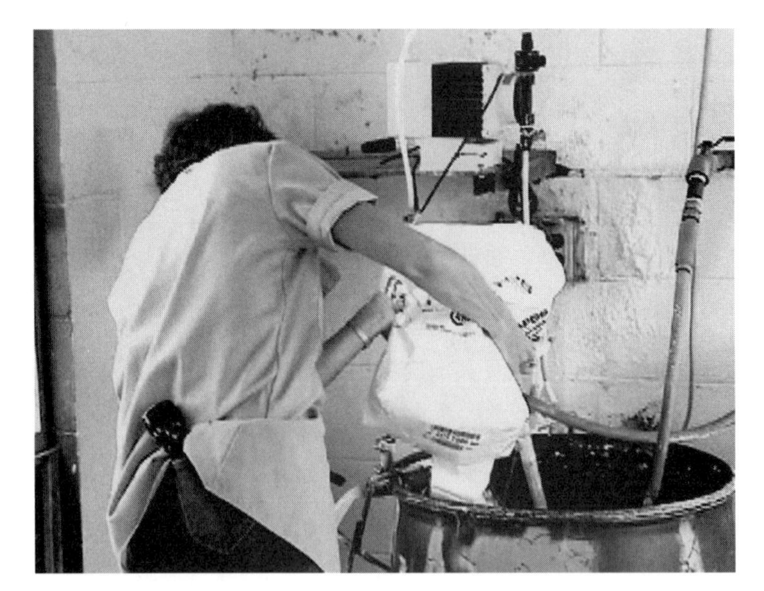

Figure C-15 A 50 lb. bag of polyphosphate "glass" is added to the solution feed barrel daily to prepare a solution of dispersant.

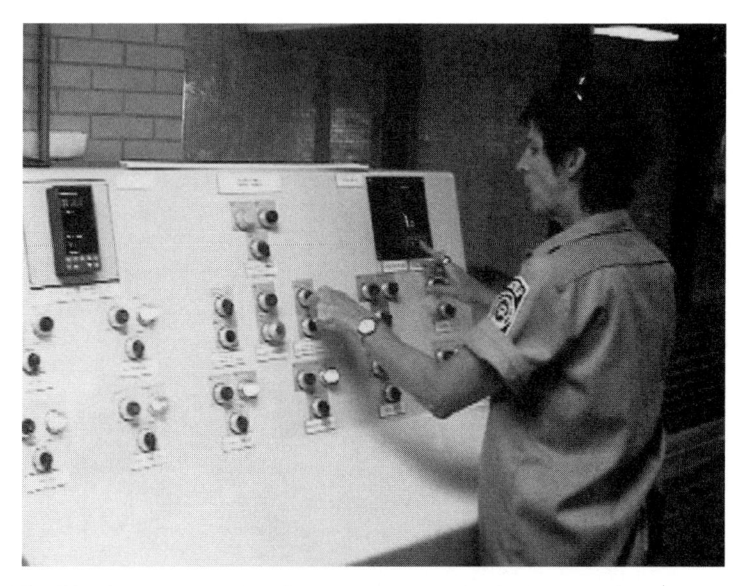

Figure C-16 Filter backwash requires filter draining, opening backwash valves, activating pump #1 for initial bed expansion, then pump #2 for further expansion.

Figure C-17 Visual observation of clarity of backwash water is followed by the successive shutdown of pumps (in order to restratify the filter media), the closing of the backwash valve, and return of the filter to service.

Figure C-18 Sludge pits are checked for depth of stored slurry.

Figure C-19 Operator samples recarbonated water.

Figure C-20 All softeners require effluent sampling plus checks of lime feed valves, flow rates, and stability of slurry blankets.

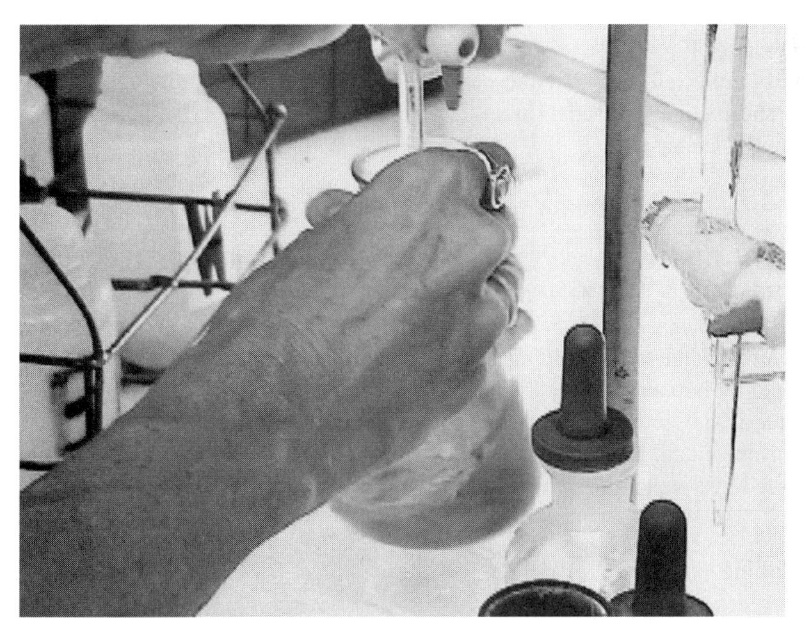

Figure C-21 Samples are analyzed for hardness and alkalinity.

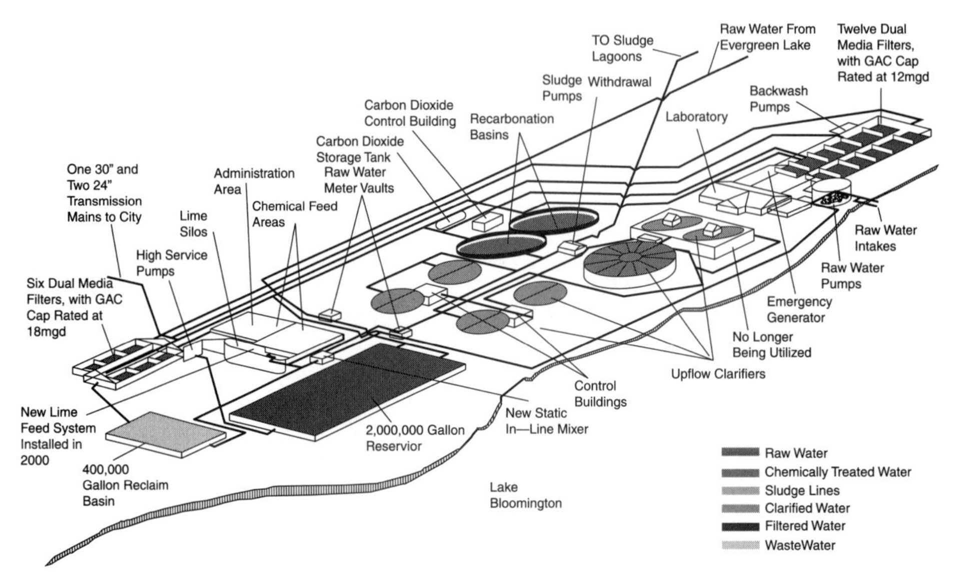

Figure C-22 Bloomington Water Treatment Plant flow diagram—2003.

polyphosphate, anionic and cationic polymers), compressors, softening basins, and recarbonation system. During the rounds, samples are collected for analysis in the operator's laboratory. While individual operators follow different paths and procedures, a typical route is demonstrated in Figs. C-4 to C-21. The entire plant campus is outlined in the flow diagram of Fig. C-22.

Samples collected during rounds are analyzed for turbidity, alkalinity, pH, chlorine, fluoride, and hardness. Results, first hand recorded on daily sheets, are later entered into the computer database.

THE LIME DELIVERY SYSTEM: FROM DRY STORAGE TO SLAKERS TO CLARICONES

A large portion of the Main Building is devoted to the lime system (Fig. C-23). Powdered lime is delivered by hopper-bottomed truck (Fig. C-24) to bulk storage tanks (Fig. C-25). The stored lime is subsequently transferred hydraulically by blowers (Fig. C-26) to a smaller dry lime storage tank located above the batch lime slakers (Fig. C-27). The lime inventory is monitored by a senior operator who maintains control over all bulk chemical supplies.

Lime Slaking and Storage

In the batch slaker, a weighed quantity of powdered lime is fed successively to each of two slakers (Fig. C-27) from the lime feed hopper. This is followed by the addition of a weighed quantity of water. Thereafter, the slaker contents are mixed and the temperature of the resulting slurry is monitored until an optimum temperature is achieved.

Figure C-23 Lime storage building.

Upon completion of the slaking reaction, the contents of the batch slaker are discharged to lime slurry storage tanks located under the floor of the slakers (Fig. C-28). The progress of the slaking process can be followed (and modified) on the operator's SCADA system.

Occasionally, lime blockages may occur at the lime feed into the slaker or at the exit pipe to the slurry storage tanks (stainless steel). To avoid process upset, the operator is responsible for detecting such blockage at the earliest possible time.

Figure C-24 Powdered lime is pneumatically transferred into building-enclosed hoppers from a hopper-bottomed truck.

Figure C-25 Lime storage hoppers.

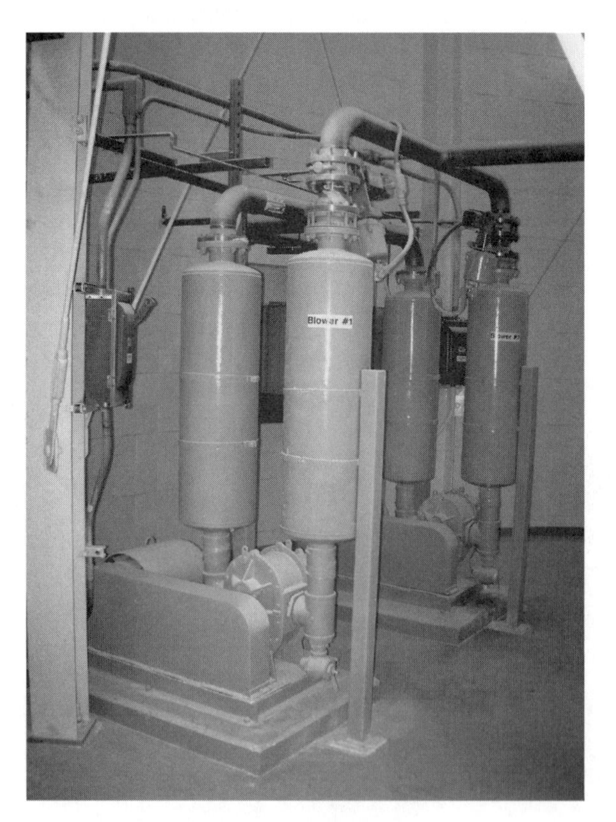

Figure C-26 Blowers are used to transfer powdered lime from bulk storage to a feed hopper atop slakers.

Figure C-27 Batch-process lime slaker.

Figure C-28 Slaked lime storage and recirculation system.

Once in storage, the hot lime slurry must be kept mixed to maintain consistency. It is constantly recirculated by pumps through flexible (black) hose in a continuous loop that enters and circles the ClariCone building. The unused lime slurry is returned to the storage tanks.

Lime Recirculation Loop

A major feature of the lime system is the lime recirculation loop through which lime slurry is delivered to each of the softening units (Figs. C-29 and C-30) and the excess flow returned to the lime storage units. This system aids in providing mixing to maintain uniformity of the lime slurry feed. However, since the lime recirculation loop suffers from occasional blockage as well as abrasion of its flexible pipe, each element requires regular attention from operators. An alternate lime feed mechanism has been established in case of prolonged blockage. Moreover, spare flexible pipe sections are available to facilitate rapid replacement.

The lime recirculation loop (flexible hose and PVC pipe, Fig. C-31) delivers slaked lime slurry to each of the four softeners. The slurry flow is controlled by pinch valves (Fig. C-32) and metered by a controller (Fig. C-33). Lime is delivered by a flexible pipe directly descending into each softening unit to a depth just above the tangential inflow of the water to be softened (Fig. C-34). After addition to the inflow, most of the slaked lime dissolves rapidly, within minutes, precipitating calcium carbonate and the more flocculant magnesium hydroxide, thereby forming a stable sludge blanket.

Figure C-29 Lime slurry feed lines entering and leaving softener building.

Figure C-30 Lime recirculation system is hung from the ceiling.

Figure C-31 Valved PVC pipes deliver lime slurry to each softening unit.

Figure C-32 Pinch valve for control of lime slurry flow.

Figure C-33 Flow meter and controller.

Figure C-34 Flexible pipe delivers lime slurry into softener.

Grit Removal

The heaviest portion (5% to 8%) of the lime slurry, consisting of sandy grit and unreactive (dead-burned) lime, settles into the bottom cylinder of the softener (approximately 1,200 gallon capacity) where it is periodically removed by grit pumps (Fig. C-35). However, before the grit pumps are activated, a tangential flow of water, called *jetting*, is used to loosen and resuspend the settled solids for more complete removal (Fig. C-36).

Figure C-35 Pump for removal of grit from base of softener.

Figure C-36 Piping for jetting and grit removal.

Grit is generally pumped from each tank in operation every eight hours for a maximum period of five minutes. It is withdrawn through an 8 inch pipe using a 1000 gpm grit pump. At this flow rate, prolonged *gritting* may result in the removal of a large portion of the slurry blanket.

If an excessive amount of heavy material (unreactive lime, calcium carbonate, sand, silt, chert, and debris from influent water) is allowed to accumulate in the bottom of the tank, it can disrupt the helical flow of influent water, directing the inflow upward, and causing short-circuiting. This may cause *upwellings* in the blanket and result in high effluent turbidities from the softener.

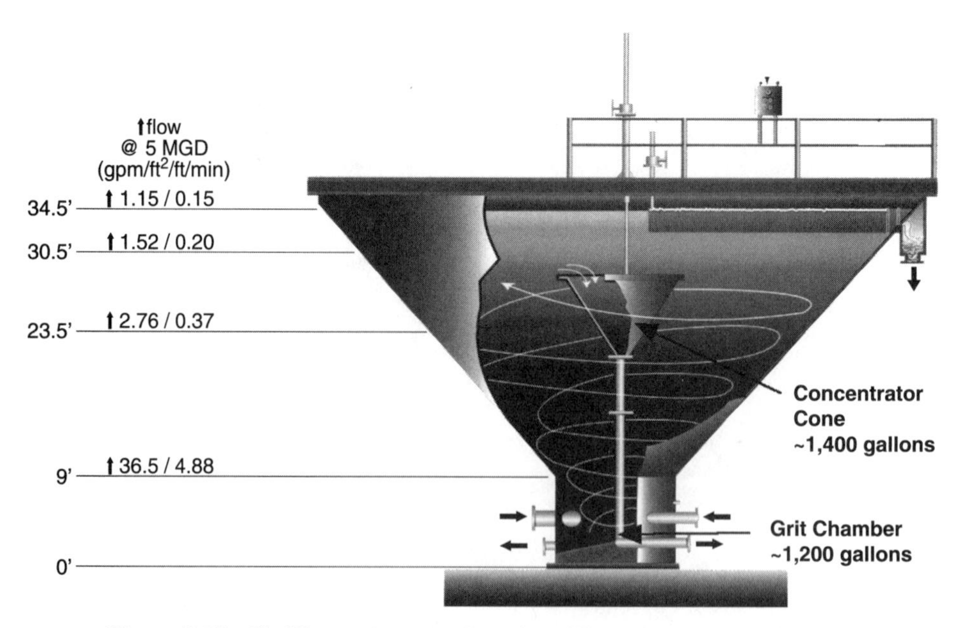

Figure C-37 ClariCone softener configuration. (Illustration courtesy of CB&I.)

Lime overfeed, as indicated by transient increases in pH of the softened and settled water (softener effluent), is observed primarily during periods when gritting takes place. Since the lime slurry feed continues uniformly, even while a portion of the softener influent water is withdrawn from the drain as part of the gritting process, the effective lime softening dosage increases.

Softener Configuration

Influent to the ClariCone softener enters through two pipes arranged tangentially to impart a spiral flow to the water in the cone-shaped tank (Fig. C-37). With ever-increasing cross-section as the flow rises, the upward flow velocity progressively decreases. Only the heaviest particles settle into the grit chamber. The lighter, flocculent particles in the slurry blanket are removed through the *concentrator cone*.

RECARBONATION: FROM LIQUID CO_2 TO GAS TO SOLUTION

Stabilization

Stabilization is accomplished by the addition of carbon dioxide to both recarbonation basins to lower the pH of the lime-treated water, thereby terminating the continued precipitation of calcium carbonate. This moderates the turbidity of water applied to the filters. Carbon dioxide is stored as a liquid in a horizontal tank, then heated and fed as a gas through two gas flow control rotameters (Fig. C-38) in the CO_2 control building.

The recarbonation basins are 60 feet in diameter, 14.8 feet deep (Fig. C-39), and have serpentine effluent launders with V-notch weirs (Fig. C-40). At the effluent channel, sodium hexametaphosphate (a dispersant) is mixed with the recarbonation basin *effluent* to prevent further aggregation of precipitates (Fig. C-41).

If an excess of lime (pH > 11) has been fed into the softening basins, dissolved calcium hydroxide in the clarified effluent will react with the carbon dioxide to precipitate additional calcium carbonate in the recarbonation basins. This post-precipitation is visualized as

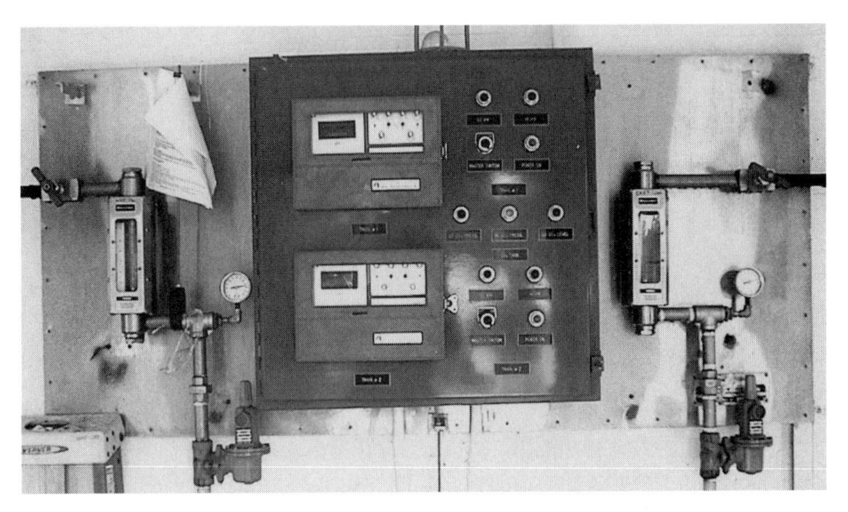

Figure C-38 CO_2 gas flow control rotameters.

Figure C-39 Influent receiving CO_2 gas.

Figure C-40 Serpentine effluent channel.

Figure C-41 Recarbonation basin—clear.

an increase in turbidity in the recarbonation basins (Fig. C-42). Operators observe the clarity of the recarbonation basin water to avoid *lime overfeed*.

Since a heater failure on the CO_2 storage tank will allow the recarbonation system to freeze, an emergency sulfuric acid storage and feed system can be used to provide for pH control and stabilization.

Figure C-42 Recarbonation basin—cloudy.

FILTRATION: DESCRIPTION AND OPERATION

Filter Sizes, Media Depths, and Flow Rates

Effective filtration is strongly dependent on previous steps in the treatment process. Ineffective pretreatment, including coagulation and sedimentation, or continuing post-precipitation of calcium carbonate, can result in elevated filtrate turbidity. As indicated by 90% reductions upon acidification, filter influent water turbidity may be almost entirely due to calcium carbonate.

There are 18 dual media (GAC/sand) filters installed at the Bloomington plant. The 12 filters in the "old" plant were constructed in three groups of four that went into service in 1929, 1956, and 1966, respectively (Fig. C-43). Each filter box contains 19 inches of granular activated carbon (GAC) overlying a 12 inch layer of 0.5 mm silica sand for a total media depth of 31 inches or 0.8 m. Surface wash is accomplished through a network of fixed nozzles suspended over the GAC layer (Fig. C-44). Used primarily for taste and odor control, the biologically-active GAC is replaced with new, virgin carbon on a three-year schedule. Four filters are serviced each year.

While rated at a surface loading rate of 1.5 gpm/ft^2 (a *downward flow velocity* of 3.7 m per hour), the 1929 plant filters are operated at only 0.6 gpm/ft^2 (1.5 m/h) This results in a substantial *empty bed contact time* (EBCT) of 19 minutes within the 19 inch (0.48 m) layer of GAC.

The six filters in the Main Building were constructed in 1994 (Fig. C-45). Each of these filters consists of two separate filter boxes served by a single influent and wash water gullet. A 24 inch (0.61 m) layer of GAC overlies a 12 inch (0.3 m) layer of filter sand (Fig. C-46). Rated at 3 gpm/ft^2 (7.3 m/h), these filters are normally operated around 2 gpm/ft^2 (4.9 m/h). For 24 inches (0.61 m) of GAC, the EBCT is 7.5 minutes. This GAC is replaced every two years.

Bloomington's filters are most convenient to operate and maintain. They are all enclosed and protected from wind-blown debris and sunlight. Individual filter banks are further enclosed within partitions with large windows for ready observation. Filter gallery areas are tiled for ease of cleaning and routine filter maintenance operations.

Figure C-43 Backwash of filter in old plant.

Figure C-44 Drained filter showing fixed surface wash nozzles.

Filter Characteristics and Performance

Filter runs are limited to 48 hours before being backwashed to avoid irreversible compaction (adhesion) of deposits on the filter media. Backwash sequences are initiated manually in order to give operators a chance to directly observe individual filters. While observing a filter backwash, Bloomington operators look for:

- *Air bubbles*, a sign of filter air binding or entrainment of air in the backwash water line
- *Calcium carbonate plates and chips*, derived from spalling or pressure washing of filter walls and piping

Figure C-45 Drained filter in the Main Building showing rotary surface wash.

Figure C-46 GAC cap during removal for replacement.

- *Foreign matter*, for example, mudballs, cemented media, algal filaments or mats, fibers, and surface accumulations
- *Uneven distribution of washwater*: boils, horizontal flows, lifting of media, wall effects, shrinkage cracks, separation of GAC and sand, and hydraulic surges
- *Media blowoff*, carryover of light media (GAC, fine sand) into the backwash effluent launders
- *Unusual color or quantities of solids* in backwash water
- *Foam*, an indication of the presence of organic (surface-tension-lowering) compounds

Bloomington operators closely monitor filter performance, particularly, in ways that computers cannot. Automated equipment, alarms, and computers have minimal powers of observation, lack judgment, and are unable to respond to emergencies. They are susceptible to many types of failure, such as electrical power surges and outages plus system component failure. Corrosion and the build-up of solids in sampling lines also cause failure of monitoring systems and the recording of erroneous data due to blockage and solids sloughing.

Each individual filter is equipped with a continuous-flow turbidimeter. In addition to SCADA data acquisition systems for recording flow and head loss, turbidity is displayed visually for each individual filter.

IEPA Turbidity Requirements By law, filtered water turbidities on any given filter should:

- Not exceed 1 ntu*
- Not exceed 0.5 ntu* after the first four hours of filter operation following a backwash

**for two consecutive fifteen minute sampling intervals.*

What to Do about Turbidity Exceedances If a particular filter is about to exceed IEPA turbidity requirements, operators take the filter out of service until the automated results can be verified and the situation properly remedied. As soon as possible, a sample of filter effluent is collected manually for analysis with the calibrated plant operator's laboratory turbidimeter to compare with the reading shown by the continuous flow turbidimeter installed on the filter.

Backwash Procedures

Accumulated particles are removed from the filters through *backwashing* with finished water. The backwash water, along with the solids removed from the filter media, can be sent to a reclaim (settling) basin. The softened supernatant water is subsequently returned to the head of the plant for reclamation.

Filter Washing Protocol Following the closing of the influent valve, lowering of the water surface level in the bed, and closing of the effluent valve and opening of the drain valve, a low-rate backwash is initiated with one backwash pump to expand the media above the level of the surface wash. Thereafter, the surface wash is activated and operated until any surface caking or accumulations are broken up.

The low-rate wash that will cause a 10% expansion of the sand layer and expand the upper bed of GAC into the rotating arms of the *surface wash* should then be conducted for approximately four minutes. (It may be desirable to minimize this wash since extended surface wash may contribute to the break-up of the friable GAC granules, thereby creating micrometer-sized carbon *fines*. Subsequently, these carbon fines may penetrate the filter following restart.)

A high-rate backwash at a nominal rate of 20 gpm/ft^2 for six to seven minutes may then be used to further expand the sand layer into the 20% to 50% expansion range. (At this point, the danger of washing out the lightest fraction of the GAC is greatest so that, based on observation, the maximum backwash rate may be limited to that which precludes the loss of GAC. The duration of the high-rate wash will be determined by the operator for the seasonal (temperature) and pretreatment conditions. For example, following the addition of polymeric coagulant aids, the backwash may be extended to assist in removing the more tightly attached polymer.)

Finally, a two-minute low-rate wash may be employed at the end of the backwash procedure to ensure gentle restratification of the filter media. Each step is detailed on the SCADA control panel.

Return-to-Service On refilling the filter to its operating (production) level, a preprogrammed, graduated *return-to-service*, or *flow ramping*, program can be initiated to bring the filter back online gradually and minimize effluent turbidity. Current filtration practice does not include flow ramping because no significant turbidity excursions are being encountered at the low rates of hydraulic filter loading currently employed at Bloomington.

GAC Replacement

Every year, under contract with Calgon Corporation, some of the GAC caps on the Bloomington plant filters are replaced with virgin (not previously used) carbon. Using Bloomington plant staff labor, the two-year-old carbon at the new plant and three-year-old

Figure C-47 GAC in totes.

Figure C-48 Replacement sand in tote.

carbon at the old plant is removed from each of the beds hydraulically. The carbon/finished water slurry is ejected from the bed through flexible hoses into an empty tractor-trailer capable of holding the GAC contents of a single filter bed. The trailer returns the used carbon to the manufacturer for thermal reactivation after which it is reused for less critical applications, such as decolorizing sugar or rum or for waste treatment processes.

Replacement GAC arrives in 1,000 pound totes that hold 40 cubic feet of the adsorbent (Fig. C-47). The bulk (dry) density of the GAC is 25 lbs/cu. ft. or 25% that of filter sand (Fig. C-48). This less dense material stratifies in filter beds during backwash to form a discrete layer of GAC atop the filter sand.

GAC Size Distribution and Characteristics After being in use for two years, the size distribution of the GAC narrows considerably. Abrasion causes larger GAC granules (>1.8 mm) to become reduced in size. Smaller GAC granules (<1 mm) appear to wash out of the filter during backwash. Overall, the narrowing size distribution with age makes the GAC, progressively, a more uniform and desirable filter medium. Normally intensely black, the carbon granules begin to lighten in shade. When dislodged from the carbon by shaking, the turbid supernatant from these granules are found to contain large numbers of microorganisms of various sizes and shapes. This is consistent with the characterization of GAC as a *biological filter medium*, since the organic nutrient adsorbed by the carbon is converted to microbial cell mass. This cell mass is part of the turbidity removed by backwash. Where biological processes are involved, a perfectly clean (organism-free) filter/adsorbent surface is not desirable.

Removal of TOC by Adsorption and Microbial Activity Total organic carbon removal on fresh GAC, initially 60%, diminishes over months as the transient *adsorptive* capacity of the virgin carbon becomes exhausted. Carbon that has been in service for one year exhibits far less TOC removal, consistent with *microbial* TOC removal alone. Overall, Bloomington plant TOC removals are far higher than the 35% required under the USEPA surface water treatment regulations. Most of the TOC reduction, about 60%, occurs during *pretreatment* (coagulation, lime softening, and sedimentation). Subsequent filtration through freshly installed GAC further increases overall TOC removal to 77%.

INDEX

Water Treatment Plant Performance Evaluations and Operations. By John T. O'Connor, Tom O'Connor, and Rick Twait